INTRODUCTION TO NEUROMORPHIC COMPUTING

Artificial intelligence (AI) is transforming industries and society, but its high energy demands challenge global sustainability goals. Biological intelligence, in contrast, offers both good performance and exceptional energy efficiency. Neuromorphic computing, a growing field inspired by the structure and function of the brain, aims to create energy-efficient algorithms and hardware by integrating insights from biology, physics, computer science, and electrical engineering. This concise and accessible book delves into the principles, mechanisms, and properties of neuromorphic systems. It opens with a primer on biological intelligence, describing learning mechanisms in both simple and complex organisms, then turns to the application of these principles and mechanisms in the development of artificial synapses and neurons, circuits, and architectures. The text also delves into neuromorphic algorithm design and the unique challenges faced by algorithmic researchers working in this area. The book concludes with a selection of practice problems, with solutions available to instructors online.

SHRIRAM RAMANATHAN is Rodkin-Weintraub Chair in Engineering at Rutgers University's College of Engineering. He previously held faculty positions at Purdue University and Harvard University and was a research staff member at Components Research, Intel. Shriram's research focuses on adaptive semiconductors for neuromorphic computing and AI, driving innovation at the intersection of materials science and next-generation AI technologies. He currently teaches the pioneering course "Semiconductors for AI" at Rutgers.

ABHRONIL SENGUPTA is Associate Professor in the School of Electrical Engineering and Computer Science at Penn State University, where he holds the prestigious Joseph R. and Janice M. Monkowski Career Development Professorship. As the director of the Neuromorphic Computing Lab, his research bridges hardware and software, focusing on sensors, devices, circuits, systems, and algorithms to enable low-power, event-driven cognitive intelligence. Abhronil also teaches the cutting-edge course "Neuromorphic Computing" at Penn State.

INTRODUCTION TO NEUROMORPHIC COMPUTING

SHRIRAM RAMANATHAN
Rutgers University

ABHRONIL SENGUPTA
Pennsylvania State University

Shaftesbury Road, Cambridge CB2 8EA, United Kingdom

One Liberty Plaza, 20th Floor, New York, NY 10006, USA

477 Williamstown Road, Port Melbourne, VIC 3207, Australia

314–321, 3rd Floor, Plot 3, Splendor Forum, Jasola District Centre, New Delhi – 110025, India

103 Penang Road, #05–06/07, Visioncrest Commercial, Singapore 238467

Cambridge University Press is part of Cambridge University Press & Assessment, a department of the University of Cambridge.

We share the University's mission to contribute to society through the pursuit of education, learning and research at the highest international levels of excellence.

www.cambridge.org
Information on this title: www.cambridge.org/9781009564342

DOI: 10.1017/9781009564335

© Shriram Ramanathan and Abhronil Sengupta 2026

This publication is in copyright. Subject to statutory exception and to the provisions of relevant collective licensing agreements, no reproduction of any part may take place without the written permission of Cambridge University Press & Assessment.

When citing this work, please include a reference to the DOI 10.1017/9781009564335

First published 2026

A catalogue record for this publication is available from the British Library

Library of Congress Cataloging-in-Publication Data
Names: Ramanathan, Shriram author | Sengupta, Abhronil author
Title: Introduction to neuromorphic computing / Shriram Ramanathan, Abhronil Sengupta.
Description: Cambridge ; New York, NY : Cambridge University Press, 2026. | Includes bibliographical references and index.
Identifiers: LCCN 2025018998 (print) | LCCN 2025018999 (ebook) | ISBN 9781009564342 hardback | ISBN 9781009564335 ebook
Subjects: LCSH: Neuromorphics
Classification: LCC TA164.4 .R36 2026 (print) | LCC TA164.4 (ebook)
LC record available at https://lccn.loc.gov/2025018998
LC ebook record available at https://lccn.loc.gov/2025018999

ISBN 978-1-009-56434-2 Hardback

Cambridge University Press & Assessment has no responsibility for the persistence or accuracy of URLs for external or third-party internet websites referred to in this publication and does not guarantee that any content on such websites is, or will remain, accurate or appropriate.

For EU product safety concerns, contact us at Calle de José Abascal, 56, 1°, 28003 Madrid, Spain, or email eugpsr@cambridge.org

Vinu, Shlok, and Sammie
Shriram Ramanathan

Family
Abhronil Sengupta

Contents

Preface		*page* xi
1	**Intelligence in Biological Systems**	1
	1.1 Intelligence and Memory in Nonneural Organisms	1
	1.2 Intelligence and Memory in Organisms with Nervous Systems	2
	1.3 Excitatory and Inhibitory Neural Network Motifs	8
	1.4 Learning Mechanisms in Simple and Complex Organisms	9
	1.5 The Quest for AI Hardware and Algorithms Inspired by Biological Intelligence	12
	References	14
2	**Principles of Artificial Neural Networks**	15
	2.1 Computational Primitives	15
	2.2 Supervised Learning	15
	2.3 Binary Classification Problem	16
	2.4 Gradient Descent	17
	2.5 Neural Network Representation	18
	2.6 Neuron Activation Functions	19
	2.7 Backpropagation	20
	2.8 Convolutional and Residual Networks	21
	2.9 Regularization	24
	References	25
3	**Artificial Synapses**	27
	3.1 Fundamental Principles of Synapse Design and Synaptic Action	27
	3.2 Filamentary Synapses	33
	3.3 Ferroelectric Synapses	38

	3.4	Insulator–Metal Transition-Based Synapses	46
	3.5	Organic Materials-Based Synapses	54
	3.6	Two-Dimensional and Layered Material Synapses	60
	3.7	Spintronic Synapses	65
	3.8	Photonic Synapses	70
		References	71
4	Artificial Neurons		77
	4.1	Fundamental Principles of Neuron Design and Neuronal Action	77
	4.2	Filamentary Neurons	81
	4.3	Ferroelectric Neurons	87
	4.4	Insulator–Metal Transition-Based Neurons	91
	4.5	Organic Materials-Based Neurons	98
	4.6	Two-Dimensional and Layered Material Neurons	102
	4.7	Spintronic Neurons	105
	4.8	Photonic Neurons	108
		References	109
5	Examples of Applications in Artificial Neural Networks		113
	5.1	Silicon CMOS-Based Neural Networks	113
	5.2	Correlated Electron Semiconductor-Based Neural Networks	115
	5.3	Filamentary Switch-Based Neural Networks	122
	5.4	Organic Electronic Neural Networks	123
	5.5	Spintronic Neural Networks	126
	5.6	Photonic Neural Networks	130
		References	131
6	System Design		135
	6.1	Neuromorphic Architecture Design	135
	6.2	Material and Device Design Impact on Circuit–Architecture Formulations	136
	6.3	Robust System Design Combatting Hardware Non-idealities	137
	6.4	Self-Healing Systems	139
	6.5	Architecture–Application–Sensing Codesign	141
		References	143
7	Neuromorphic Algorithms		145
	7.1	Spiking Neural Network Training Methodologies	145
	7.2	Local Learning	151
	7.3	Probabilistic Computing	153

	7.4 Dynamic Networks	155
	7.5 Energy-Based Models and Learning	158
	7.6 Hybrid Algorithm Designs Leveraging Conventional Deep Learning Aspects	162
	7.7 Need for Hardware–Algorithm–Application–Neuroscience Codesign	163
	References	164
8	Lifelong Learning with AI Algorithms and Hardware	169
	8.1 Emulating Complex Neural Functions beyond Plasticity	169
	8.2 Adaptation, Learning, and Evolutionary Dynamics	172
	8.3 Bioinspired versus Bio-realistic Functions in AI Algorithms and Hardware	175
	8.4 Future Directions in Biomimetic Computation	177
	References	180
9	Practice Problems	182
	References	188
Index		189

Preface

Neuromorphic computing is a branch of artificial intelligence (AI) that seeks to emulate biological intelligence in hardware and algorithms toward the ambitious goal of low power electronics. It is a highly interdisciplinary research and technology area that brings together physics, chemistry, engineering, computer science, psychology, and neuroscience. Given the breadth of the field, it is rather challenging indeed to strike a balance between establishing the underlying principles within each technical domain and its intersection with neuromorphic computing when preparing a textbook. This is what we have attempted to do in this book, introducing the key fundamental topics related to neuromorphic computing starting from semiconductor materials and devices to circuits and algorithms. In addition, practical considerations such as variability and its impact on AI hardware and algorithm design are discussed. Fundamental concepts of neuroscience that have both historically guided and continue to impact the field of neuromorphic computing are also discussed.

At the time of writing the book, there is incredible interest across academia and industry to implement AI in various application domains from physics to health. This has led to exploration of various neuromorphic architectures mostly implemented in software and running on traditional silicon circuits. In this scenario, one must worry about whether this growth is sustainable. The power consumption of electronic devices and AI-related technology domains is now approaching that of the total energy budget of many countries. Clearly, low power electronics is highly desirable and urgently needed for a sustainable future. One approach to enable this is building computers that are inspired by the brain. Through evolution, the brain has figured out to perform complex tasks such as facial recognition, ensuring survival in dynamic environments, and so on, with limited power consumption. How can we then emulate the remarkable information processing properties of the brain in computing devices and circuits and make them operate with ultralow power consumption? Concurrently, AI systems continue to face multiple algorithmic challenges like catastrophic forgetting, adversarial susceptibility, requirement of huge amounts of labeled training data, and lack of uncertainty quantification in confidence-critical applications, to name a few. How can we borrow principles from brain-inspired learning approaches that overcome such training bottlenecks? The reader will be well equipped to address these questions and identify further areas of research in this fascinating topic after reading the book.

Finally, the authors would like to express their sincere thanks to Sarah Armstrong (Editor) and Aleksandra Serocka (Content Manager) of Cambridge University Press for their guidance throughout the various stages of the production of this book.

1
Intelligence in Biological Systems

1.1 Intelligence and Memory in Nonneural Organisms

Although this book is primarily concerned with neuromorphic computing, that is, emulating intelligence found in the nervous system of various organisms; for completeness sake, we begin the discussion with a brief overview of intelligence in simple organisms. Examples of simple organisms include the slime mold, amoeba, stentor, and others that do not possess a well-defined nervous system. In recent years, a substantial body of work has been published reporting various attributes of intelligence, memory, and learning capabilities in such unicellular organisms. This is quite surprising as the typical notion of intelligence is associated with the brain and the complexity of the neuronal networks in the brain that is connected with rest of the body. The slime mold *Physarum polycephalum* (referred to as slime henceforth) is a model system that has received much attention. Slime utilizes protoplasm distribution through its body using a network of tubules. The pulsing action of veins in this network can affect the mechanical shape by creating pressure gradients and direct its motion. For instance, rhythmic oscillatory motion of the cytoskeleton in the slime close to a food source is enhanced directing the organisms toward it by pushing the protoplasm while the opposite occurs when exposed to a noxious stimulus. Essentially, this implies that coupling of mechanical oscillators across the body of the organism enables movement toward a food source or away from a danger signal. Indeed, this suggests the flow of information across the body via fluidic channels. The ability of slime to solve maze puzzles as well as create efficient networks for finding optimal paths are well-known examples of complex problem-solving ability [1]. Interestingly, despite not having a brain that can store memories, the slime is able to navigate toward food sources without continuously going around in circles. This behavior is enabled by the slime utilizing external spatial memory: The organism leaves behind extracellular slime as it navigates a path and can sense its presence, thereby avoiding traversing these same regions [2]. More interestingly, presence of a food source allows the slime to overrule this characteristic. In other words, a hierarchy of decision-making skills can be found in slime. In recent years, biologists have studied the mechanisms of learning in slime. For example, habituation, a fundamental form of learning, has been observed in slime. Habituation is the process of an organism learning to ignore a non-harmful stimulus when presented repeatedly. When different strands of genetically compatible slime are connected, the learning characteristics can also be transferred, hence information is contained in the protoplasm about the environment which can be shared. Such remarkable abilities of the amoeba noted in various studies including short-term memory, decision-making, and event anticipation have led to development of circuit models using a combination of inductors, capacitors, and resistors with memory to emulate the learning features [3]. We point to the reader this is one of the approaches to translating information processing in unicellular organisms into algorithms, devices, and circuit-level analogies. That is, adapting knowledge from biology into design of electrical circuits and algorithms that emulate some functional aspects of organismic behavior. Much of this book is dedicated to developing algorithm and hardware building blocks that can emulate aspects of intelligence in more complex organisms that have a nervous system.

1.2 Intelligence and Memory in Organisms with Nervous Systems

From here on, we focus on the biological underpinnings of intelligence in organisms that possess a nervous system and a brain. Indeed, neuromorphic computing aims to emulate intelligence arising from animal brains. Hence it is worth a brief discussion of the fundamentals of natural intelligence accessible to engineers (very much like the authors!) who do not have formal training in neuroscience. This section is certainly not meant to be comprehensive, but rather a succinct summary of key aspects of brain matter, particularly, neurons and synapses and their interconnections that drive information processing and memory. We refer the reader to neuroscience literature that discusses the nuances in detail [4]. Let us begin with a brief discussion on synapses that are considered responsible for learning and memory. Synapses are interconnections between neuronal cells and govern information flow in neural circuits. They can be simply thought of as wires that connect neurons with the additional feature that their resistance is not a fixed quantity preset by their geometry. Synapses can change their efficacy (or weight) to transfer signals between neurons based on experience. In other words, experience-dependent modification of the weight is a central characteristic of synapses and is referred to as synaptic plasticity. For instance, if a synapse receives stimulus from two

1.2 Intelligence and Memory in Neural Organisms

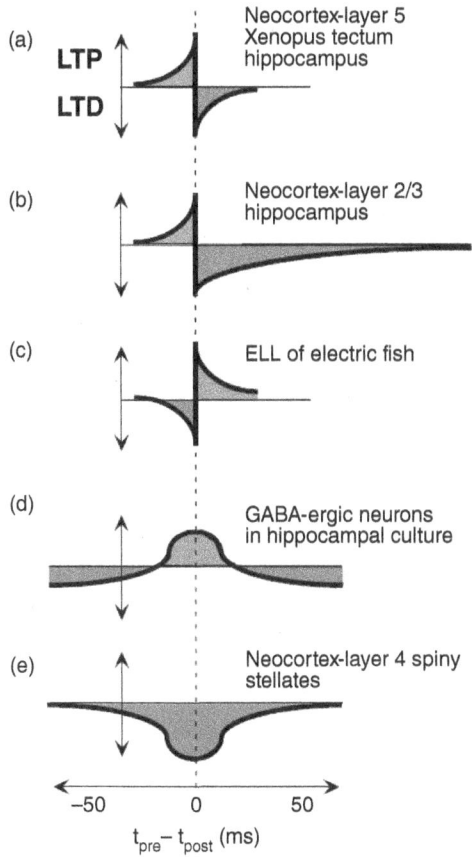

Figure 1.1 Synaptic plasticity in various preparations is shown. In (a–c), the sequence of presynaptic versus postsynaptic spike determines the potentiation or depression characteristics, in (d and e) only the relative timing of the spikes matter but not their sequence [5]. Adapted with permission from [5]. Copyright © 1969, Nature America Inc.

neurons (referred to as presynaptic and postsynaptic) within a short time interval say of the order of few tens of milliseconds, the synapse can change its weight. If the two stimuli are received across larger time intervals, then there is no change to the weight, and the synapse does not consider these events correlated. This is referred to as spike timing-dependent plasticity (STDP) and is widely studied not just in neuroscience but also in neuromorphic computing. Figure 1.1(a–e) summarizes some of the plasticity mechanisms noted in neuroscience [5]. Hebbian plasticity refers to the fact that repeated stimulation of a neuronal cell "B" by cell "A" results in strengthening the connection (i.e., synapse weight) between them. This is one of the classic models for plasticity postulated by Hebb in the 1940s. The pre- and postsynaptic spike sequence may or may not affect the potentiation and depression characteristics depending on the neuronal pathways. It is important to note however that there are many forms of synaptic plasticity that occur over a range of timescales. Note that additional stimuli in the form of neuromodulators (NMs) can also perturb the plasticity. A synapse can be primed for plasticity via a slowly decaying signal called eligibility trace (ET). Neuromodulators that occur in delayed time can trigger the plasticity of such selective synapses. Hence, there may be a larger timescale involved that exceeds what is typically considered in Hebbian STDP. Supervised synaptic plasticity occurs when the output is not absolute but rather the difference between actual value and an instructional pattern. The instructive pathway arises from a different region in the brain and the signal is transmitted through the dendritic structure of the neurons. Hence, these mechanisms are more complex and distinct from simple Hebbian correlation of input–output. Behavioral timescale synaptic plasticity (BTSP) combines several of these features resulting in a longer timescale for activity of a synapse.

In the biological setting, the plasticity of synapses is modulated by various biochemical matter such as neurotransmitters and proteins released in the neuronal circuits. Same argument holds for short-term (seconds timescale) versus long-term memory (hours, weeks, or through the life span). Hence, there are characteristic diffusional and reactional timescales involved that control the plasticity across regions while local activity can drive additional changes. We will discuss interactions between synapses in Chapter 8. Synapses can be broadly classified into two types, namely electrical and chemical. Electrical synapses represent direct connection between neurons via gap junctions that are intercellular channel clusters. In chemical synapses, neurotransmitters released at one end of a neuron are detected by receptors in another neuron. Due to the different conduction mechanisms, electrical synapses are faster (milliseconds), while chemical synapses are slower (seconds) but can amplify signals. The two types of synapses coexist in the brain and can even interact [6]. Each type of synapse is valuable for a specific function in the brain. For instance, electrical synapses can be bidirectional and their rapid speed of signaling is useful for an animal to respond to active threats. Chemical synapses on the other hand are slower but capable of amplifying signals that are important for learning. Neuromodulators can affect the function of electrical synapses and induce plasticity. It is important to note that the gap junctions and chemical synaptic connections are not only plastic but also can form and disappear during the growth stages of an organism. Hence, their presence and function are tied to the developmental stages of the nervous system and interactions between them. Chemical synapses are probabilistic in nature while electrical synapses are not. That is, the ejection of neurotransmitters occurs in a probabilistic manner. Plasticity in electrical synapses is possible due to changes in number of cellular channels at the gap junctions and is therefore a dynamic unit. Of course, in chemical synapses, NMs (e.g., dopamine and serotonin) can profoundly modify

1.2 Intelligence and Memory in Neural Organisms

their weights. Studies on the rat brain have shown that mixed synapses comprising electrical and chemical receptors can be present. Having both types of synapses in proximity therefore enables a multitude of approaches to shunt currents to or away from neurons to control the excitability in a circuit over distinct timescales. There is a growing body of work that suggests that synapses are not only crucial for memory but also control the information flow in neural circuits and therefore directly participate in neural computation. We will revisit the finer complexity of neural circuit components toward the latter chapters of this book.

Next, we discuss neurons that are responsible for the generation of electrical signals (referred to as action potentials) that correspond to information flow in neural pathways. Neurons comprise a cell body referred to as soma from which numerous sub-components referred to as neurites emerge. These include dendrites that receive information from surroundings and transmit to the cell body, and an axon that conducts information to different locations. There are multiple types of neurons such as multipolar neurons, bipolar neurons, pseudo-unipolar neurons, and unipolar neurons. The primary differences between these types are the relative positions of the cell body and the dendrites and presynaptic terminals at the end of the axon. The different neuron types are distributed across different parts of the nervous system such as in the brain and retina. Neurons can further be classified into two groups referred to as Golgi type I and Golgi type II. The former possess long axons, while the latter have very short axons and are interneurons. The functional characteristics of the neurons depend on their location in the nervous system and are influenced by the presence and generation of neurotransmitters that can excite or inhibit their activation probability. Glial cells are present in the nervous system that serve many functions including providing electrical isolation between neurons, serving as a barrier for neurotransmitters and can even enable transport of ions using their networks [7]. Neurons are encased by a neuronal membrane that has ion channels embedded in them. There exists a potential difference across the cell membrane referred to as resting potential. Ion channels can be non-gated or gated. In the former, ionic flow is permitted during the membrane resting period. The gated channels can also open and close in response to changes in membrane potential. When the channel is closed, it is in resting state, and active when it is open. The channel also can be inactivated following an active period and is referred to as refractory during this interval.

Various ionic species are constantly diffusing in and out of neuronal membranes in response to stimuli and due to concentration gradients. Further, cell membranes also contain various biochemical species that can have fixed charges (e.g., anions) that induce potential gradients across the cell membrane. Na^+, K^+, Cl^-, and Ca^{2+} are some of the ions responsible for generation of action potentials. Typically, there exist unequal concentration of ions inside and outside of the membrane due to different number of binding sites for each ion. This results in a potential difference determined by the Nernst equation in the ideal case considering activity of just one ionic species. For instance, for potassium ions, the equilibrium potential is predicted to be about −80 mV, while in measurements, it is closer to −65 mV due to other ions such as Na^+ that can leak in and out of the membrane. Note that this balance is due to a delicate interaction between the concentration gradient of ions that tend to push K^+ out of the membrane while electrostatic interaction with fixed negative charge attracts it back into the membrane. At the same time, Na^+ ions from outside the membrane face a similar pressure to maintain their balance and interact with K^+ ions in this process. Note that the number of channels that allow Na^+ permeation versus K^+ are not necessarily equal and this reinforces the concentration gradient across the membrane. Hence, the neuronal membrane components play a critical role in maintaining ionic equilibria. The Goldman–Hodgkin–Katz (GHK) equation presents one approach to estimate the resting potential considering multiple ionic species that enter and exit the neuronal membrane. The GHK equation is given by

$$V = \frac{RT}{F} \ln \frac{P_K [K^+]_{out} + P_{Na} [Na^+]_{out} + P_{Cl} [Cl^-]_{in}}{P_K [K^+]_{in} + P_{Na} [Na^+]_{in} + P_{Cl} [Cl^-]_{out}}, \quad (1.1)$$

where P_K, P_{Na}, and P_{Cl} are the membrane permeabilities for the K^+, Na^+, and Cl^- ions, respectively. Note that if ionic species that are not monovalent are also involved in determining the resting potential, Equation 1.1 must be suitably modified.

So how is an action potential generated? We can address this question by considering what happens when a neuron receives an excitatory input. Ion channels that are permeable to Na^+ open resulting in massive influx of sodium ions. The membrane potential is therefore altered and starts increasing. When it reaches a threshold potential, there is 50% chance the neuron will fire an action potential. And when this is exceeded, the probability approaches 1. To balance this sudden increase in Na^+ influx, K^+ ion channels open with a slight time delay resulting in outward diffusion. This process begins to reset the polarization state of the neuron back to the resting potential and is known as the falling phase. At the end of this phase, the Na^+ ion channel becomes inactive for a brief period known as refractory period. Beyond this timescale, the process can begin again depending on the input stimulus strength. Once the action potential is generated near an axon hillock, it is propagated down the axon by passive spread of current. The exact current distribution also depends on whether neurons have an insulation sheath of myelin around them. In each myelinated neuron, there exist regions referred to as Nodes of Ranvier that have voltage-gated Na^+ channels that assist in periodic regeneration

1.2 Intelligence and Memory in Neural Organisms

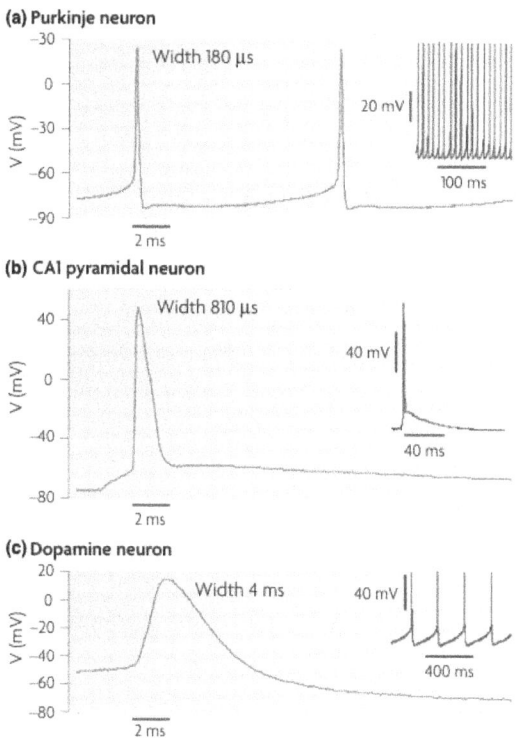

Figure 1.2 (a–c) Recorded action potentials from different neurons in the mammalian brain. Note the diversity in peak shape and temporal width of the action potential. The rate of repolarization of the neuronal cell membrane dictates the sharpness of the voltage spike [8]. Adapted with permission from [8]. Copyright © 1969, Springer Nature Limited.

of the signal. The propagation of the action potential is also referred to as saltatory conduction. Figure 1.2 shows an example of action potential recorded from neurons in a mouse brain [8].

A wide variety of neuronal action potentials have been noted in experimental measurements on mammalian brain slices. The width of the voltage spike and frequency of spiking are dependent on the neuron type. Perhaps this is not surprising, since the shapes of the voltage curves will depend on the local dynamics of ionic transport across cell membranes and establishment of equilibrium conditions until the arrival of the next stimulus. It is therefore very clear that all neurons are not alike in signal generation even within a single animal body. As we will discuss in Chapter 8, contemporary neuroscience research delves deeper into the internal structure of the neurons and how different compartments within a neuron can interact to produce action potentials. For instance, how does having complex dendritic spines in a neuron affect its computational capacity and so forth? Neuromorphic

1.3 Excitatory and Inhibitory Neural Network Motifs

Neurons can enable other neurons in their network to fire an action potential or limit their activity. The former are referred to as excitatory while the latter is inhibitory in nature. In a neural network, there are continual electrochemical interactions between neurons through their synapses and surrounding glial cells. Depending on the stimulus received from the presynaptic terminal, a neuron can be cajoled into

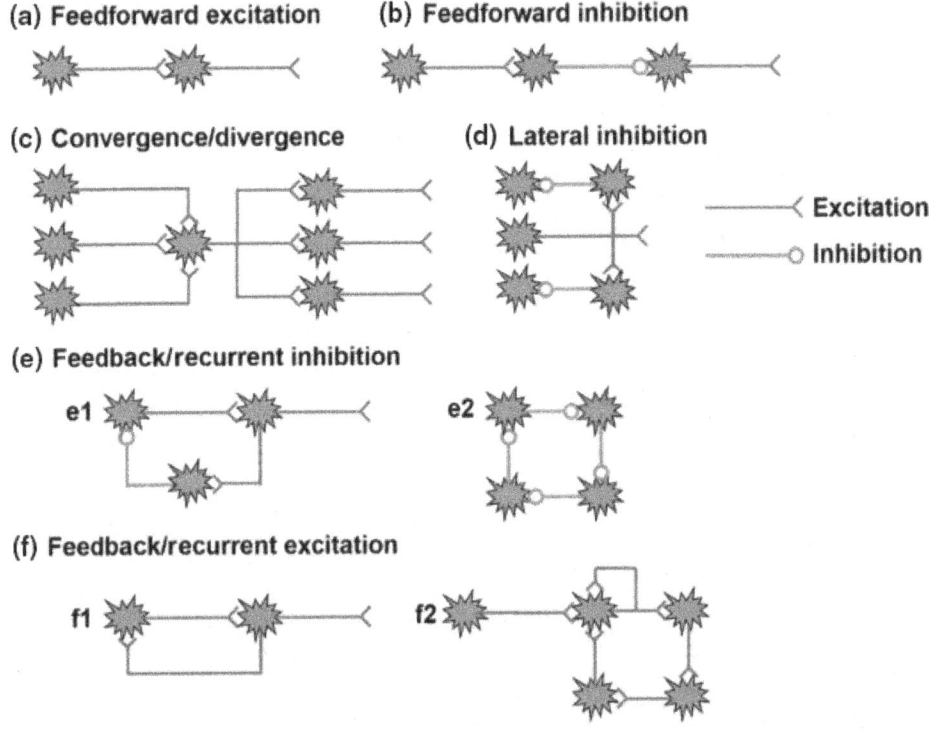

Figure 1.3 (a–f) show different types of connections that can exist between neurons. The line with circle connections represents inhibitory connections, while the line with less than connections represents excitatory connections. Excitatory connections enhance the probability of neuronal action potential generation while inhibitory connections reduce it. Note that neurons can give feedback to the presynaptic cells in addition to transmitting action potentials to adjacent neurons. This feedback mechanism can serve as a control knob for maintaining neural activity in a circuit [9]. Adapted with permission from [9]. © 2022 The Authors. Advanced Materials published by Wiley-VCH GmbH.

firing an action potential that can propagate further down the axon. On the other hand, an inhibitory neurotransmitter can reduce the voltage gradient across the cell membrane suppressing the membrane potential from reaching threshold. Action potential from a presynaptic neuron that depolarizes the membrane of the postsynaptic cell is referred to as excitatory postsynaptic potential (EPSP). The term "excitatory" refers to the fact that the membrane potential is pushed closer to the threshold value needed to generate an action potential. On the other hand, a presynaptic neuron that releases neurotransmitters results in hyperpolarization of the postsynaptic cell membrane making it more negative and thereby inhibits the firing probability. This type of potential is referred to as inhibitory postsynaptic potential (IPSP). Note that single presynaptic pulses are often insufficient to generate action potential in a postsynaptic neuron. Therefore, sequences of potentials that are entering a postsynaptic cell add up to result in a net potential that is compared against the threshold resulting in generation of an action potential. There is summation both in space (at dendritic regions) and time (interval between various signals) occurring.

Neurons can receive and transmit signals among just a few or up to several thousands of neurons in a network. The complexity of the networks can be broken down into simple building blocks to understand the fundamental processes. Figure 1.3 shows examples of network microcircuits that are present in the brain [9]. The diversity in connections between neurons is responsible for numerous applications including edge detection, establishing biological rhythms such as circadian clock cycles via gene regulation that control sleep/wake patterns, maintaining balance and posture during locomotion.

1.4 Learning Mechanisms in Simple and Complex Organisms

It is now well established that even simple unicellular organisms can learn from their environment. Learning is of course central to our lives and closely associated with intelligence. For instance, we often talk about learning from our mistakes to avoid costly errors in the future. Of course, this requires us to store the memory of the events surrounding the original mistake somewhere in our brain! In neuromorphic computing, one primary goal is to design machines that can continuously learn. Therefore, we need to understand the fundamental mechanisms behind learning in organisms and examine which of these aspects can be abstracted and implemented in hardware and / or at the algorithmic level in computing function. Organisms that do not have a nervous system still display remarkable properties essential for survival. The simplest and perhaps universal form of learning is non-associative. Habituation and sensitization represent two types of learning among this category. Habituation is the process of an organism demonstrating reduced response to repeated stimuli while sensitization is the opposite, a noxious stimulus evokes a sharp response. Slime molds have been shown to display habituation to caffeine while searching for food. What elementary processes are involved in this learning trait? It has been postulated that movement of protoplasm inside the organism can be controlled by oscillatory networks of fluid flow which can adapt to environmental cues. Habituation has also been studied in the ciliate *Stentor*. The *Stentor* contracts when a mechanical stimulus is applied. Repeated application of such a stimulus result in weakened contraction while stronger stimulus generates a new response. The novel response in turn indicates the weakening of the contraction is not simply due to fatigue. The rate of habituation also closely depends on the strength of the stimulus demonstrating unique response to the perturbation. Mechanoreceptor cells in *Stentor* respond to the stimulus cue by generating action potentials due to calcium ion migration across the cell membrane. It has been postulated that ion channel properties are modified during the stimulus application resulting in changes to the conductance of the membrane [10]. Cellular mechanisms therefore can control the response of an organism despite not having a central nervous system to make decisions. Researchers have therefore labeled this as a primitive form of intelligence that may have been present in organisms even before the evolution of the brain.

Learning in organisms that possess a nervous system is more complex since multiple neural circuits could be involved. Understanding the molecular basis of learning across various animal species is a major research theme in neuroscience. At the same time, one must consider the various forms of learning ubiquitous in such organisms that go beyond non-associative learning [11]. *Aplysia* has been widely studied as a model system to understand the mechanisms of learning and memory in neuroscience. We discuss some key results to present the role of neurons and synapses in learning and memory formation that serve as inspiration for artificial intelligence (AI) algorithms and hardware.

Let us now consider a form of learning that requires more than one stimulus. Associative learning involves forming connections between two stimuli and learning to respond to cues. Classical conditioning and operant conditioning are two forms of associative learning. The former involves pairing a neutral stimulus with a stimulus that elicits a response. Perhaps a famous example is the Pavlovian dog that was trained to salivate when a bell was rung. This feat was accomplished by constantly pairing food and sound of a ringing bell. Hence, an initially neutral stimulus namely a ringing bell can cause salivation in a dog by pairing with food. On the other hand, operant conditioning requires an organism to modify its response based on behavior. Presentation of a stimulus therefore depends on the specific response and hence a reward can modify the probability of the response over repeated trials. Fear conditioning studies in various animals are also conducted in a similar manner to observe their behavior such as freezing upon

1.4 Learning Mechanisms in Simple and Complex Organisms

presenting a shock stimulus. Non-associative learning that requires only a single stimulus can be simply categorized into habituation, sensitization, and dishabituation. Habituation refers to an organism decreasing its response to repeated presentation of non-harmful stimuli, while dishabituation is the process of recovery of habituated behavior when a strong stimulus is presented. Sensitization is quite the opposite, that is an amplification of response to a noxious stimulus. It is interesting to note that the sensitization response of an animal lasts longer when the intensity of the noxious stimulus gets higher.

When we are discussing learning mechanisms, we are inadvertently discussing the collective response of an ensemble of neural components that results in a macroscopic observation, such as salivation of a dog to a ringing bell or withdrawal of the gill of *Aplysia* subject to electric shocks. So, to make the connection between the elementary components and the macroscopic behavior, we need to know what the basic neural circuits are that are responsible for the specific behavior and how they are modified during the learning process. This also begs the question: How long does an animal remember the stimulus and its original response?

Extensive studies have been carried out on *Aplysia* to understand the molecular mechanisms involved in memory. For instance, application of electric shock to the siphon results in withdrawal of the gill (respiratory organ). Repetitive application of the stimulus leads to a reduction in the withdrawal while a strong stimulus can cause sensitization. The effects of a single sensitizing shock lasts for minutes while prolonged shocks can lead to weeks long memory effect. The sensory neurons in *Aplysia* are connected to the motor neurons via synapses and this circuit is profoundly affected by the electric stimuli. For instance, studies have shown that NMs such as serotonin control the synaptic response in these circuits. Protein generation following the stimulus can modify the excitatory nature of the synapses. In other words, activity-dependent neuromodulation is responsible for the macroscopic observations involving the withdrawal of the gill. Hence, short-term adaptive behavior and long-term memory retention are linked by the microscopic phenomena such as generation and distribution of neurotransmitters, synthesis of new proteins and their subsequent effects on neural pathways.

Similar conclusions have been made based on studies on habituation response in *C. elegans*. *C. elegans* is a model organism in neuroscience owing to its well-studied nervous system that contains 302 neurons. Habituation learning of mechanical stimulus has been shown to occur due to modification of chemical synapses at the interface between sensory neurons and command interneurons. Studies on various other organisms have shown that modification of synaptic connections can indeed result in learning environmental cues and stimuli. In other words, we can conclude that in organisms possessing a nervous system, synaptic wiring modification; concurrent changes to action potentials in neurons and their propagation characteristics are important for learning and memory. Combined, this represents manifestation of intelligence in outward behavior. Beyond the elementary learning behavior described earlier that largely concern sensory response, for more complex forms of memory such as in the hippocampus, a combination of synaptic junctions can store memory in a shared manner [12]. Sharing the memory states among synaptic junctions enables fault tolerance and can operate in noisy environments. These networks are also described as auto-association networks and synthetic analogs for neuromorphic computing will be discussed in Chapters 5–7.

1.5 The Quest for AI Hardware and Algorithms Inspired by Biological Intelligence

The biological brain and nervous system combined represent a set of machinery that routinely performs complex computation and makes decisions. Decisions involve wide ranging areas from enhancing survival chances in harsh environments, optimal approaches to finding food sources, and ensuring species propagation via

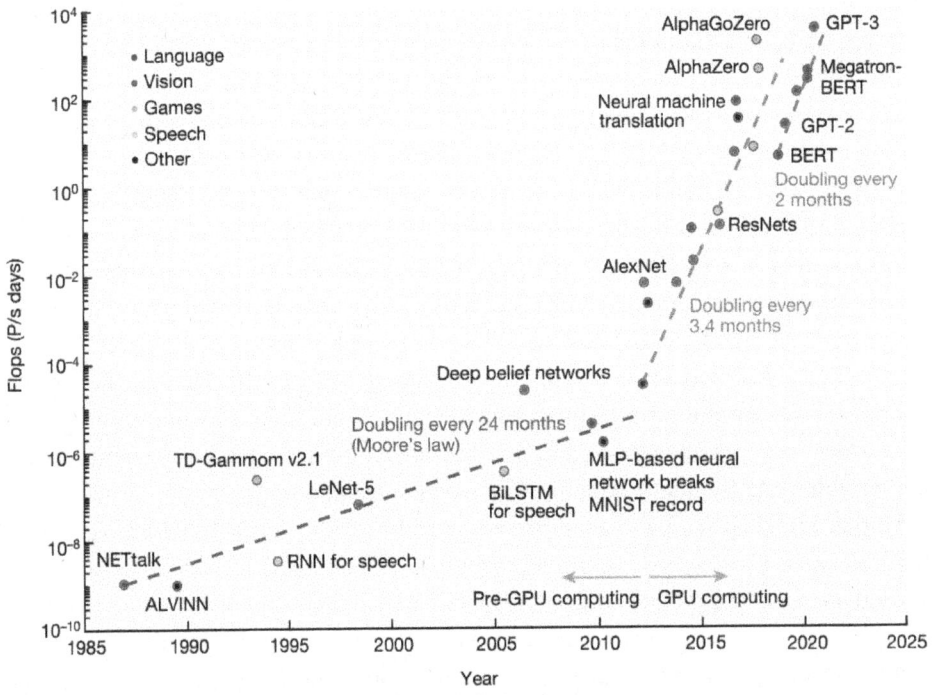

Figure 1.4 Rapid growth of computational requirements to enable AI workloads. While the computational requirements in the pre-GPU computing phase were in line with Moore's law, the current era of large language models is seeing a computational requirement increment of almost 12-fold faster than Moore's law [14]. Adapted with permission from [14]. © 2022. Springer Nature Limited.

reproductive success. A remarkable aspect of the brain is the inherently low power consumption. According to most estimates, the human brain consumes about 20 W of power. This is an unprecedentedly low number compared to the energy hungry computers designed by humans that require orders of magnitude greater power for performing comparable tasks such as facial recognition or responding to simple commands. A recent study reported that training common deep learning models can result in a carbon footprint equivalent to five times the lifetime emission of an average car [13]. Another study has outlined the progressive increment of AI compute requirements over the years (see Figure 1.4) [14]. In the pre-GPU computing era, the floating-point operations per second (FLOPS) requirement doubled every two years, which followed transistor scaling rate (known as Moore's law). However, the GPU-based computing phase has seen a FLOPS increment of 2× every 3.4 months. With the advent of large language models (LLMs), the 2× increment is occurring every 2 months – a rate that is 12 times faster than Moore's law. Hence, a fundamental rethinking is required for AI hardware with innovations stemming from semiconductor materials and devices to circuits and architectures that mimic the parallel, in situ computing capabilities of the brain. Nature has had billions of years to evolve complex organs and biochemical circuits for ensuring survival of each species amidst very cluttered and noisy environments. The question is therefore whether we can learn from nature to design machines that can operate at low power, are fault tolerant, can dynamically learn during their lifespan, and adapt to new knowledge to suitably modify future decisions. Algorithms loosely based on concepts from animal intelligence are already widely used in the software field. These include the use of artificial neural networks where nodes represent neurons and the interconnections between neurons represent synapses with tunable weights. An input is processed by a layer of neurons and the results of the calculations are readout. Multiple layers of neurons can be added as the problems get more complex. To train the network, synaptic weights are modified by rules such as backpropagation, which will be discussed in Chapter 2. However, such computational paradigms lack bio-fidelity and do not mimic the sparse, temporal event-based processing of neurons, local learning of synapses, and self-repair capability of glial cells, to name a few. AI algorithms today have forayed far from the bio-inspiration route and driven more by statistical and mathematical improvements to the black-box optimization problem of training a neural network to solve a particular task. They also suffer from a host of algorithmic challenges such as adversarial susceptibility, requirement of huge amount of labeled data during training, and lack of adaptation capability to changing environments. Thus, there is a critical need to revisit AI algorithms from a brain-inspiration route. Neuromorphic computing is centered exactly around these goals to realize robust, efficient, adaptive, self-healing, and low-power computing primitives for emerging AI technologies [15].

References

[1] Smith-Ferguson, J. and Beekman, M., 2020. Who needs a brain? Slime moulds, behavioural ecology and minimal cognition. *Adaptive Behavior, 28*(6), pp. 465–478.

[2] Reid, C. R., Latty, T., Dussutour, A. and Beekman, M., 2012. Slime mold uses an externalized spatial "memory" to navigate in complex environments. *Proceedings of the National Academy of Sciences, 109*(43), pp. 17490–17494.

[3] Pershin, Y. V., La Fontaine, S. and Di Ventra, M., 2009. Memristive model of amoeba learning. *Physical Review E – Statistical, Nonlinear, and Soft Matter Physics, 80*(2), p. 021926.

[4] Kandel, E. R., Schwartz, J. H., Jessell, T. M., Siegelbaum, S., Hudspeth, A. J. and Mack, S. eds., 2000. *Principles of neural science* (Vol. 4, pp. 1227–1246). New York: McGraw-Hill.

[5] Abbott, L. F. and Nelson, S. B., 2000. Synaptic plasticity: Taming the beast. *Nature Neuroscience, 3*(11), pp. 1178–1183.

[6] Pereda, A. E., 2014. Electrical synapses and their functional interactions with chemical synapses. *Nature Reviews Neuroscience, 15*(4), pp. 250–263.

[7] Siegel, A. and Sapru, H. N., 2006. *Essential neuroscience.* Lippincott Williams & Wilkins.

[8] Bean, B. P., 2007. The action potential in mammalian central neurons. *Nature Reviews Neuroscience, 8*(6), pp. 451–465.

[9] Park, T. J., Deng, S., Manna, S., Islam, A. N., Yu, H., Yuan, Y., Fong, D. D., Chubykin, A. A., Sengupta, A., Sankaranarayanan, S. K. and Ramanathan, S., 2023. Complex oxides for brain-inspired computing: A review. *Advanced Materials, 35*(37), p. 2203352.

[10] Wood, D. C., 1988. Habituation in Stentor: A response-dependent process. *Journal of Neuroscience, 8*(7), pp. 2248–2253.

[11] Byrne, J. H., Heidelberger, R. and Waxham, M. N. eds., 2014. *From molecules to networks: An introduction to cellular and molecular neuroscience.* Academic Press.

[12] Rolls, E. T., 2013. A quantitative theory of the functions of the hippocampal CA3 network in memory. *Frontiers in Cellular Neuroscience, 7*, p. 98.

[13] Strubell, E., Ganesh, A. and McCallum, A., 2020, April. Energy and policy considerations for modern deep learning research. In *Proceedings of the AAAI Conference on Artificial Intelligence* (Vol. 34, No. 9, pp. 13693–13696).

[14] Boahen, K., 2022. Dendrocentric learning for synthetic intelligence. *Nature, 612*(7938), pp. 43–50.

[15] Mead, C., 1990. Neuromorphic electronic systems. *Proceedings of the IEEE, 78*(10), pp. 1629–1636.

2
Principles of Artificial Neural Networks

2.1 Computational Primitives

Let us consider an example problem concerning predicting the price of a car based on the number of people it can accommodate, the car's model, the year of manufacturing, and the location of the dealership selling the car, among other factors. However, there may be other hidden factors as well. For instance, the car's model might involve other aspects like trim or edition builds which can impact the price. Price ranges may also fluctuate based on the location of the dealership itself (country, state, and district). Therefore, often times, it is difficult to handcraft a network architecture consisting of decision nodes that might fit a particular problem solution. The great advantage of a neural network is that it can figure out such hidden relationships automatically. We will go into details of the neural network architecture in Section 2.5. Before proceeding, it is important to highlight supervised learning and a simple binary classification problem to build up to the context of neural networks.

2.2 Supervised Learning

A significant section of machine learning problems currently relies on supervised learning wherein we have large amounts of labeled data that can be used for training a neural network model. For instance, regular neural networks can be used in problems like car price prediction. Various computer vision problems such as object recognition and detection rely on convolutional neural networks (CNNs) [1]. The huge popularity of large language models (LLMs) in the past few years is attributed to specific architectures such as bidirectional encoder representations from transformers (BERT) [2] and generative pre-trained transformer (GPT) [3] that uses a specific module termed as "attention" [4]. Confidence critical applications like self-driving cars use probabilistic neural networks like Bayesian nets [5] that can provide an additional uncertainty metric as the output of the neural network that can aid in the decision-making process. However, reducing the amount of labeled data during training is an important research problem as one cannot expect a human annotator to provide labels in real-time in edge-intelligence scenarios (for instance, running a deep learning model on a resource-constrained mobile phone). Therefore, unsupervised and few-shot learning modes are being actively looked at by the research community.

Supervised learning using neural networks can deal with structured (like predicting price of a house) as well as unstructured data (like image or video processing). The success of deep learning is attributed to the fact that the performance (for instance, classification accuracy in an image classification problem) gradually improves as the amount of labeled data is increased during the training process. While conventional pattern recognition tools such as logistic regression and support vector machines may even outperform neural networks for small amount of labeled data, their performance saturates off after a certain limit of labeled data is reached during the training process. The success and popularity of deep learning are due to its scale, which have not only been made possible by algorithmic advancements at training large neural networks with billions of parameters but also by significant advancements in the computing hardware itself and availability of large amounts of data in the era of Internet of Things.

2.3 Binary Classification Problem

Let us consider a binary classification problem with an input image x, and we have to classify whether the image is that of a boat. Let us represent this decision by a binary variable y that assumes the value of 1 corresponding to a boat and 0 otherwise. As a particular example, let us consider that x is an image of size 256×256 pixels with red (R)–green (G)–blue (B) channels. Therefore, the dimension of x is a vector of size $S_x = 256 \times 256 \times 3 = 196{,}608$ pixels (considering that all channel values are concatenated in the form of a vector).

Let us now frame this in the form of a logistic regression problem. We assume that we have access to a dataset where we have a training set that contains input images mapped to corresponding labels. On the other hand, the user does not have access to labels of the testing set while training the model that is used to test the trained model's performance by comparing its prediction against the true labels. To perform the prediction, we consider that each dimension of the input x is multiplied by a corresponding scaling factor that represents its importance. We represent this by the vector $w \in \mathbb{R}^{S_x}$. Let us also consider an additional bias term b, thereby resulting in the prediction of the model being represented by:

$$\hat{y} = P(y=1|x) = \sigma(w^T x + b),\tag{2.1}$$

where σ represents the nonlinear sigmoid function, $\sigma = \frac{1}{1+e^{-x}}$. The sigmoid function assists in mapping the unbounded linear function of $w^T x + b$ to a number lying between 0 and 1, thereby representing the probability of the image being that of a boat. It is worth noting here that implementation of Eq. (2.1) from a coding perspective should preserve the dot-product kernel instead of using explicit iterative "for" loops. This is because many computing platforms today, such as graphics processing units (GPUs), are optimized for matrix calculations. Here, w and b are unknown parameters that one is trying to optimize to minimize the overall prediction error over the entire training set. Let us represent the i-th training image as x^i and the corresponding label as y^i. Therefore, we would like the prediction of the network to be as close as possible to the true labels, and this can be achieved by formulating the prediction problem as a minimization of the following cost function:

$$E = \frac{1}{d}\sum_{i=1}^{d}(y^i - \hat{y}^i)^2,\tag{2.2}$$

where d is the number of training samples. However, this cost function results in a non-convex optimization problem. Therefore, to formulate this problem as a convex optimization problem, the cost function can be restructured as [6]:

$$E = \frac{1}{d}\sum_{i=1}^{d} -y^i \log \hat{y}^i - (1-y^i)\log(1-\hat{y}^i).\tag{2.3}$$

While we are not focusing on the derivation of Eq. (2.3), let us intuitively discuss the implications of the function. If $y^i = 1$, $E = \frac{1}{d}\sum_{i=1}^{d} -\log \hat{y}^i$. Minimization of this function would imply $-\log \hat{y}^i$ to be as low as possible, thereby causing $\hat{y}$ to be as high as possible, that is, 1. Similar observations can be made for the other scenario of $y^i = 0$.

2.4 Gradient Descent

To minimize the error function, we adopt the strategy of gradient descent [7]. Gradient descent is an iterative process whereby the unknown weight and bias parameters are updated sequentially depending on the magnitude and sign of the derivative of the error function. For instance, at each iteration, the weight and bias parameters are updated as,

$$w_{t+1} = w_t - \alpha \frac{\partial E(w,b)}{\partial w}; \quad b_{t+1} = b_t - \alpha \frac{\partial E(w,b)}{\partial b},\tag{2.4}$$

where w_t and b_t are weight and bias parameters at the t-th iteration. The learning rate α dictates the step size of the parameter update. A very high learning rate can result in losing the minima during the optimization process, while a low learning rate can result in slow training convergence. Gradient descent essentially causes the parameters to converge in the direction of steepest descent in the context of a convex optimization landscape. If the sign of the gradient is positive (an increment in w increases E), then Eq. (2.4) promotes the weights in the next iteration to reduce (thereby reducing E) and vice versa.

Let us now go through the process of calculating the derivatives for the i-th weight and a single input image ($d = 1$). In order to calculate Eq. (2.4), we require $\partial E / \partial w_i$ that can be calculated by decomposing the operations through the following steps:

Step 1 (Calculation of $w^T x$, where w_i and x_i are the vector components):
$$z = \sum_i w_i x_i + b.$$

Step 2 (Calculation of sigmoid nonlinearity): $\hat{y} = \sigma(z)$.

Step 3 (Calculation of error function): $E = -y\log\hat{y} - (1-y)\log(1-\hat{y})$.

Let us now proceed backward through the above steps for calculation of the gradient term:

Gradient Step 3: $\dfrac{\partial E}{\partial \hat{y}} = -\dfrac{y}{\hat{y}} + \dfrac{1-y}{1-\hat{y}}$

Gradient Step 2: $\dfrac{\partial E}{\partial z} = \dfrac{\partial E}{\partial \hat{y}} \dfrac{\partial \hat{y}}{\partial z} = \dfrac{\partial E}{\partial \hat{y}} \hat{y}(1-\hat{y}) = \hat{y} - y$

Gradient Step 1: $\dfrac{\partial E}{\partial w_i} = \dfrac{\partial E}{\partial z} \dfrac{\partial z}{\partial w_i} = \dfrac{\partial E}{\partial z} x_i = (\hat{y} - y) x_i.$

2.5 Neural Network Representation

Let us first consider a three-layer neural network as shown in Figure 2.1 where we have an input layer followed by a hidden layer and an output layer. Each node of the network is a neuron that is represented by an activation function (sigmoid nonlinearity σ in this case). Neurons are connected to each other through weights that represent the synaptic functionality. As we proceed through the feedforward equations, it is worth noting that a neural network is essentially a large number of logistic regression units arranged in parallel layers. The input layer represents the input features to the network which represent the pixel values of an input image. In vector notation, we represent the input features as x of dimension S_x. Let us consider the weight matrices joining the input–hidden and hidden–output layers as w^1 and w^2, respectively, while the corresponding bias matrices are b^1 and b^2, respectively.

2.6 Neuron Activation Functions

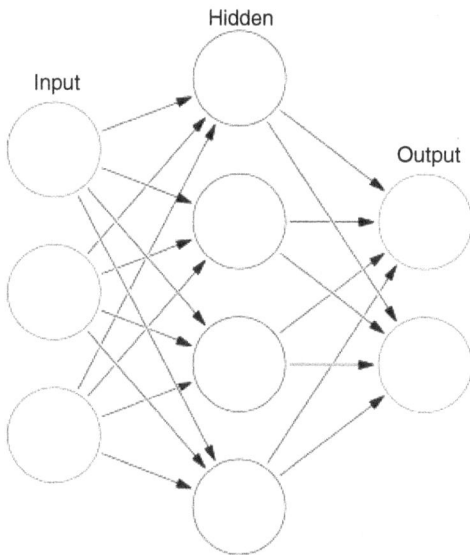

Figure 2.1 An example of a three-layered fully connected neural network with an input, a hidden, and an output layers [8]. Adapted from [8], under the Creative Commons Attribution-Share Alike 3.0 Unported license. CC BY 3.0.

In this specific example for a neural network with three input features, four hidden neurons, and two output neurons, the sizes of the various matrices are as follows: $w^1 = (4,3); w^2 = (2,4); b^1 = (4,1); b^2 = (2,1)$. The network feedforward equations are therefore given by:

$$z^1 = w^1 x + b^1; \hat{y}^1 = \sigma(z^1); z^2 = w^2 \hat{y}^1 + b^2; \hat{y}^2 = \sigma(z^2), \qquad (2.5)$$

where z^1, z^2 represent the dot-product computations of the individual layers and $\hat{y}^1, \hat{y}^2$ represent the neural activation outputs of the two layers, respectively. Equation (2.5) can be further vectorized across multiple training samples by considering them to be concatenated along columns of the matrix x.

2.6 Neuron Activation Functions

Our previous discussion has mostly focused on sigmoid activation function. However, this often encounters convergence issues during gradient backpropagation since it is a saturating function (gradient becomes zero in saturating regions of the graph). Learning improves with tan*h* function since it is zero-centered and aids in the learning process. The issues encountered with saturating functions are solved by using non-saturating nonlinearities like ReLU (rectified linear unit) [9] or its smoother counterparts like leaky ReLU [10] and Gaussian error linear unit (GELU) [11].

Next, we discuss the importance of the activation function by considering a scenario where it is set to zero. In that case, Eq. (2.5) can be simplified as:

$$z^1 = w^1 x + b^1; \hat{y}^1 = z^1; z^2 = w^2 \hat{y}^1 + b^2; \\ \hat{y}^2 = z^2 = w^2 \left(w^1 x + b^1 \right) + b^2 = w' x + b', \quad (2.6)$$

where $w' = w^2 w^1$ and $b' = w^2 b^1 + b^2$. Therefore, it is apparent that without the neural activation, neural networks would significantly lose their representation power, even if more layers are added. Therefore, nonlinear activation functions are key to the current success of deep neural networks.

2.7 Backpropagation

Determining the weights and biases of such a multilayer neural network follows a similar procedure as outlined in Section 2.4. Let us consider the three-layer neural network shown in Figure 2.1. The first step involves a forward propagation of the input sample through the layers of the network. While the derivation of backpropagation equations for such a network is less intuitive than the steps noted in Section 2.4, it is worth noting that current GPU-based deep learning libraries such as PyTorch and Tensorflow contain in-built functions that automatically calculates the gradients needed for calculation of the weights and biases. Without going into derivation details, we list out the equations for updating the parameters of the two layers as follows [6]:

$$\frac{\partial E}{\partial z^2} = \hat{y}^2 - y \quad (2.7)$$

$$\frac{\partial E}{\partial w^2} = \frac{1}{d} \frac{\partial E}{\partial z^2} \hat{y}^{1T} \quad (2.8)$$

$$\frac{\partial E}{\partial b^2} = \frac{1}{d} \text{sum}\left(\frac{\partial E}{\partial z^2} \right) \quad (2.9)$$

$$\frac{\partial E}{\partial z^1} = w^{2T} \frac{\partial E}{\partial z^2} * \hat{y}^{1\prime} \quad (2.10)$$

$$\frac{\partial E}{\partial w^1} = \frac{1}{d} \frac{\partial E}{\partial z^1} x^T \quad (2.11)$$

$$\frac{\partial E}{\partial b^1} = \frac{1}{d} \text{sum}\left(\frac{\partial E}{\partial z^1} \right) \quad (2.12)$$

Here, * represents element-wise multiplication and $\hat{y}^{1\prime}$ represents the derivative of the activation values. One can easily map the equations for the last layer of the neural network (Eqs. (2.7) and (2.8)) to the ones noted in Section 2.4. The process of training is referred to as backpropagation since the error gradients are propagated backward through the layers of the network. The neural network is usually initialized with random weights and then trained using backpropagation with the aid of other tricks like regularization that will be discussed in Section 2.9. Various hyperparameters such as learning rate and batch size of the training set used for each weight update step control the convergence of the training process and quality of the final solutions.

2.8 Convolutional and Residual Networks

Consider the previous problem of recognizing a boat from images of dimension 256 × 256 pixels with RGB components. If we use a neural network with fully connected structure and the first hidden layer containing 1,000 neurons as an example, the weight matrix for the first layer will have a dimension of $256 \times 256 \times 3 \times 1{,}000 = 1.96 \times 10^8$. However, real images have much higher dimensions, and therefore this is clearly not a sustainable solution for large-scale image recognition tasks.

The key idea behind CNNs is to replace this fully connected architecture with convolution operation, thereby resolving the dimensionality problem. A typical CNN structure (LeNet5 [12]) is shown in Figure 2.2, where the two main operators are convolution and subsampling/pooling. For instance, note the first convolutional layer C1. C1 has six convolutional filters of size 5 × 5 that results in six output feature maps of size 28 × 28. The convolutional filters are learnt by backpropagation. In general, a filter of size $f \times f$ when convolved with an input map of size $n \times n$ produces an output map of size $\lfloor \frac{n-f+2p}{s} + 1 \rfloor \times \lfloor \frac{n-f+2p}{s} + 1 \rfloor$, where p is the number of pixels used for padding the input map and s is the convolution stride. The number of convolution kernels in a layer dictates the number of output maps where each kernel is three dimensional with the third dimension corresponding

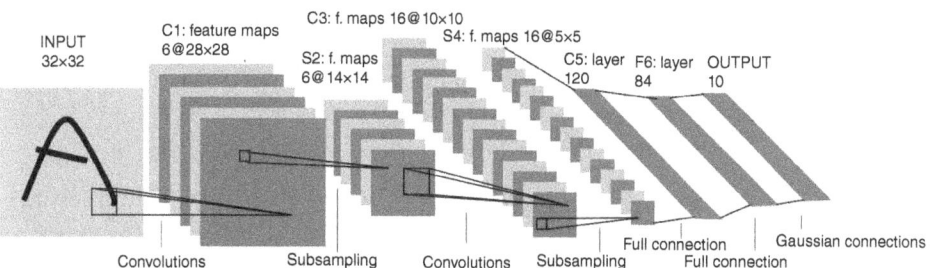

Figure 2.2 An example of CNN architecture used for digit recognition: LeNet5 [12]. Adapted with permission from [12]. Copyright © 1998, IEEE.

to the number of input maps. The convolutional output is subsequently passed through the neuron nonlinearity like ReLU.

The pooling or subsampling layer (for instance, S2 in Figure 2.2) is used for reducing the dimensionality of the convolutional output. A common choice is to subsample the input map by a factor of 2 where the map is divided into nonoverlapping regions of 2 × 2 pixels, and subsequently, the maximum or average value of the pixels in that subregion is used to update the corresponding location in the output map. Pooling operator does not involve any trainable parameters.

The CNN structure is associated with multiple advantages. First, the trainable weight map is significantly sparse as each value in the output map depends only on a small subset of the input map. As an example, the first C1 layer in the LeNet structure has only $5 \times 5 \times 3 \times 6 = 450$ trainable parameters. It also utilizes the concept of "parameter sharing" [6] through the intrinsic use of the convolution operator. Other popular CNN architectures are AlexNet [13] and VGG [14], among others. Popular benchmark computer vision datasets used for performance comparison of new algorithmic solutions include MNIST [15], CIFAR [16], and ImageNet [17].

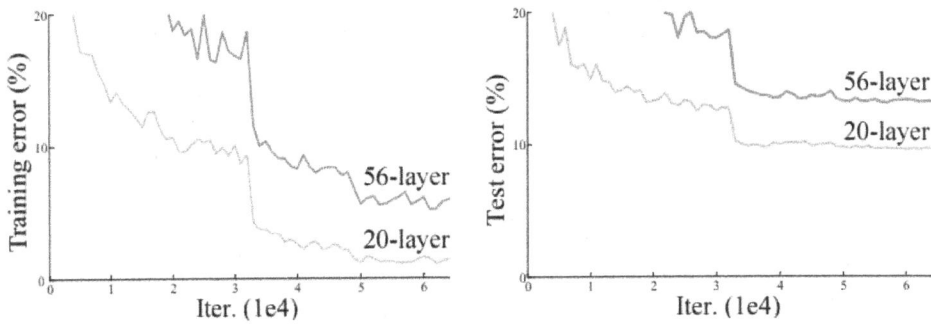

Figure 2.3 Training and test errors of 20- and 56-layer networks without residual connection [18]. Adapted with permission from [18]. Copyright © 2016, IEEE.

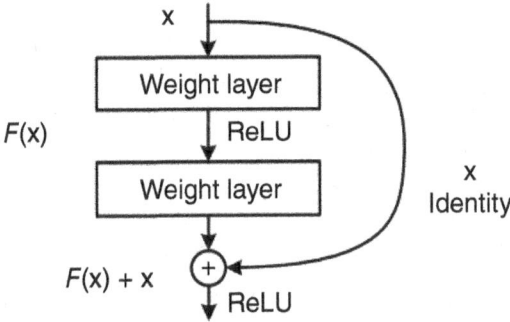

Figure 2.4 Core functional primitive of a residual network [18]. Adapted with permission from [18]. Copyright © 2016, IEEE.

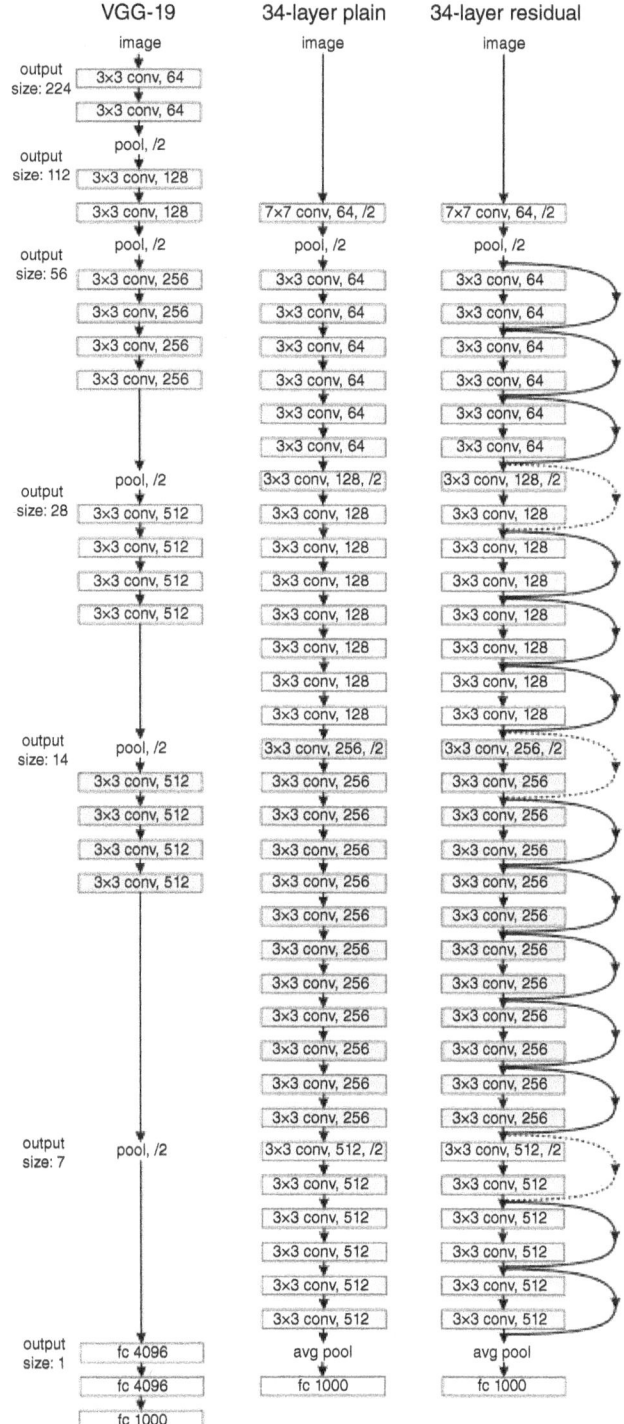

Figure 2.5 Example of network architectures used for the ImageNet dataset. The dotted shortcut connections are used to increase the dimension [18]. Adapted with permission from [18]. Copyright © 2016, IEEE.

A common trend observed in CNN architectures is that the map dimension reduces in size, while the number of maps increases with the increasing network depth.

While deep CNNs have demonstrated significant success in a variety of computer vision problems, performance usually starts saturating and degrading for very deep networks. Interestingly, this is not an overfitting problem as both training and testing errors start decreasing with increasing depth (see Figure 2.3) [18]. To circumvent this issue, He et al. [18] introduced the concept of residual networks where an identity mapping is added to the mapping introduced by the convolutional layer to simplify the optimization problem (see Figure 2.4). Figure 2.5 shows an example of a 34-layer residual network that can be used for the ImageNet dataset [18]. Residual architectures are promising enough to train networks as deep as 150 layers [18].

2.9 Regularization

The current success of deep learning is also made possible by multiple proposed concepts to aid backpropagation training. One such effect is regularization that aims to ensure that the neural network is trained to achieve the optimal decision boundary and is not overfitting or underfitting the problem. One method to achieve regularization is to add the norm of the weight matrix to the loss function mentioned in Eq. (2.3) with a scaling factor that modulates the degree of regularization. Let us consider a saturating neural transfer function like "sigmoid" to intuitively explain this concept. For very high values of the regularization term, the optimization process will cause the weights to converge to small values, thereby utilizing the linear portion of the transfer curve. This is equivalent to a large section of the neural units losing their nonlinear effects, thereby acting as a control knob to transition from overfitting to underfitting scenario. With an optimal setting of the regularization term, the optimal decision boundary can be achieved. The new loss function results in a backpropagation step update that includes an additional weight decay term (weight values from previous update step scaled by a fractional number) [6]. Another popular regularization mechanism is termed dropout. Dropout is used specifically during the training process where every neuron is assigned a particular probability of generating an output. Effectively, this causes the neural network to distribute the weights equally across all features as it cannot rely on any single feature due to the dropout effect. Therefore, this intuitively has a similar regularization effect as discussed earlier. Finally, another regularization effect frequently utilized in neural network training is termed batch normalization that aims to speed up the training process. This is inspired by the fact that inputs to a neural network are usually normalized to have zero mean and unit variance. This helps in causing the input variables to have similar number ranges, thereby preventing oscillations in the gradient descent process. However, this usually does not ensure that the inputs to deeper layers of the network will have normalized values. Therefore, one can introduce a batch normalization layer [19] after each neuronal layer, where the activation $\hat{y}^i$ is first normalized to have zero mean and unit variance:

$$\hat{y}^i_{norm} = \frac{\hat{y}^i - \mu}{\sqrt{\sigma^2 + \varepsilon}}, \tag{2.13}$$

where μ and σ are the mean and variance values of $\hat{y}^i$ and ε is a small constant. Subsequently, the normalized activation value is also scaled and shifted by parameters β and γ, thereby resulting in the batch normalization layer output $\tilde{y}^i$:

$$\tilde{y}^i = \beta \, \hat{y}^i_{norm} + \gamma. \tag{2.14}$$

The parameters β and γ are learnt by backpropagation. It is worth noting that each mini-batch is scaled by the mean and variance of the activation values of the corresponding mini-batch. Therefore, this adds some noise effect similar to dropout during the training process. As the mini-batch size increases, this regularization effect starts reducing. Much of the success of training very deep neural networks can be attributed to batch normalization [20].

References

[1] Li, Z., Liu, F., Yang, W., Peng, S. and Zhou, J., 2021. A survey of convolutional neural networks: Analysis, applications, and prospects. *IEEE Transactions on Neural Networks and Learning Systems*, *33*(12), pp. 6999–7019.

[2] Devlin, J., 2018. BERT: Pre-training of deep bidirectional transformers for language understanding. arXiv preprint arXiv:1810.04805.

[3] Achiam, J., Adler, S., Agarwal, S., Ahmad, L., Akkaya, I., Aleman, F. L., Almeida, D., Altenschmidt, J., Altman, S., Anadkat, S. and Avila, R., 2023. GPT-4 technical report. arXiv preprint arXiv:2303.08774.

[4] Vaswani, A., Shazeer, N., Parmar, N., Uszkoreit, J., Jones, L., Gomez, A. N., Kaiser, Ł. and Polosukhin, I., 2017. Attention is all you need. *Advances in Neural Information Processing Systems*, *30*, pp. 1–11.

[5] Jospin, L. V., Laga, H., Boussaid, F., Buntine, W. and Bennamoun, M., 2022. Hands-on Bayesian neural networks: A tutorial for deep learning users. *IEEE Computational Intelligence Magazine*, *17*(2), pp. 29–48.

[6] Coursera Deep Learning Specialization Course, www.coursera.org/specializations/deep-learning.

[7] Amari, S. I., 1993. Backpropagation and stochastic gradient descent method. *Neurocomputing*, *5*(4–5), pp. 185–196.

[8] Source: https://commons.wikimedia.org/wiki/File:Colored_neural_network.svg.

[9] Nair, V. and Hinton, G. E., 2010. Rectified linear units improve restricted Boltzmann machines. In *Proceedings of the 27th International Conference on Machine Learning (ICML-10)* (pp. 807–814).

[10] Maas, A. L., Hannun, A. Y. and Ng, A. Y., 2013, June. Rectifier nonlinearities improve neural network acoustic models. In *Proceedings of the ICML* (Vol. 30, No. 1, p. 3).

[11] Hendrycks, D. and Gimpel, K., 2016. Gaussian error linear units (GELUs). arXiv preprint arXiv:1606.08415.
[12] LeCun, Y., Bottou, L., Bengio, Y. and Haffner, P., 1998. Gradient-based learning applied to document recognition. *Proceedings of the IEEE, 86*(11), pp. 2278–2324.
[13] Krizhevsky, A., Sutskever, I. and Hinton, G. E., 2017. ImageNet classification with deep convolutional neural networks. *Communications of the ACM, 60*(6), pp. 84–90.
[14] Simonyan, K. and Zisserman, A., 2014. Very deep convolutional networks for large-scale image recognition. arXiv preprint arXiv:1409.1556.
[15] Deng, L., 2012. The MNIST database of handwritten digit images for machine learning research [best of the web]. *IEEE Signal Processing Magazine, 29*(6), pp. 141–142.
[16] Krizhevsky, A. and Hinton, G., 2009. Learning multiple layers of features from tiny images. www.cs.utoronto.ca/~kriz/learning-features-2009-TR.pdf.
[17] Deng, J., Dong, W., Socher, R., Li, L. J., Li, K. and Fei-Fei, L., 2009, June. ImageNet: A large-scale hierarchical image database. In *2009 IEEE Conference on Computer Vision and Pattern Recognition* (pp. 248–255). IEEE.
[18] He, K., Zhang, X., Ren, S. and Sun, J., 2016. Deep residual learning for image recognition. In *Proceedings of the IEEE Conference on Computer Vision and Pattern Recognition* (pp. 770–778).
[19] Bjorck, N., Gomes, C. P., Selman, B. and Weinberger, K. Q., 2018. Understanding batch normalization. *Advances in Neural Information Processing Systems, 31*, pp. 1–12.
[20] Ioffe, S., 2015. Batch normalization: Accelerating deep network training by reducing internal covariate shift. arXiv preprint arXiv:1502.03167.

3
Artificial Synapses

3.1 Fundamental Principles of Synapse Design and Synaptic Action

Artificial neural networks are conceptualized as layers of neurons interconnected by synapses (see Figure 3.1) [1]. As mentioned earlier, given a set of inputs provided to a neuron, each of them are multiplied by an "importance" value which represents the synaptic efficacy or weight for that corresponding input. This dot product then passes onto the neuron resulting in an output signal which drives the neuronal inputs for the next layer. Therefore, from a computational standpoint, synapses represent the memory elements while neurons represent the processing unit. The degree of detail involved in modeling such neuro-synaptic computations varies depending on the generation of artificial neural network being considered. For instance, most current deep learning algorithms abstract synapse functionality as a weight value while the neuron functionality is usually represented by a

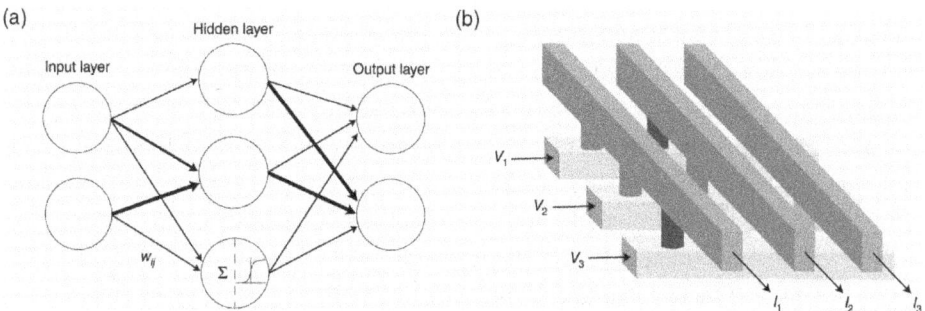

Figure 3.1 (a) The computational primitive in an artificial neural network consists of neuronal layers interconnected by synaptic weights. (b) The weighted summation of synaptic inputs followed by neuronal processing can be mapped to a crossbar array of nonvolatile synaptic devices interfaced with neuronal devices by simple application of Kirchhoff's law and Ohm's law [1]. Adapted with permission from [1]. Copyright © 2019, Springer Nature Limited.

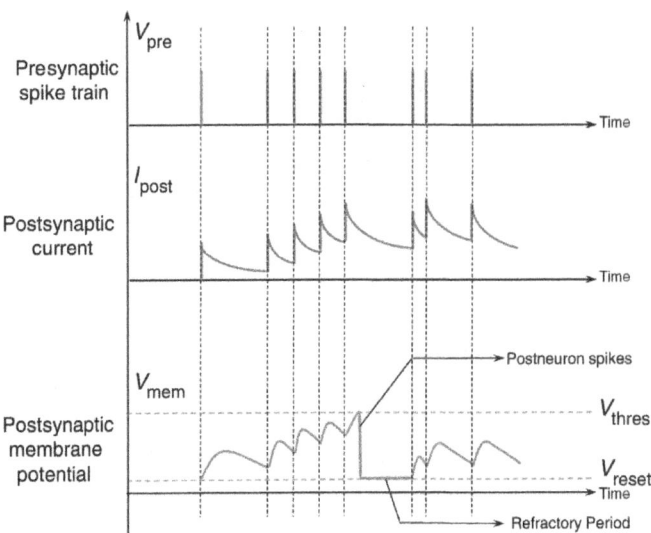

Figure 3.2 Neuro-synaptic dynamics in a spiking neural network [4]. Adapted with permission from [4]. © 2016 American Physical Society.

nonlinearity function such as sigmoid and rectified linear unit (ReLU). Such networks will be referred to as analog neural networks (ANNs) in this book due to the analog representation of neuron input/output signals and was the topic of discussion in Chapter 2. As discussed in Chapter 2, ANNs can be arranged in convolutional architectures [2] and have achieved state-of-the-art results in a large plethora of pattern recognition tasks [3].

However, biological signal propagation between neurons is temporal and driven largely by binary spike signals instead of analog values. Driven by the goal of modeling neuro-synaptic computations with a much higher degree of bio-fidelity, most neuromorphic computational models therefore abstract artificial neural networks as spiking neural networks (SNNs) where signal communication between neurons is event-driven. Figure 3.2 depicts transient waveforms of neuro-synaptic dynamics in response to a spike train [4]. Each voltage spike signal triggers a postsynaptic current response that increments by the synaptic strength value at every spike and decays exponentially otherwise. This postsynaptic current is integrated subsequently by the neuron and reflected in the membrane potential dynamics. The neuron membrane potential is also characterized by leak dynamics. Once the membrane potential crosses a particular threshold, an output spike is generated. Given such a binary model of signal propagation and computation, SNNs have been highly attractive for the AI hardware community due to its promise of enabling sparse, event-driven computation and communication. Intuitively, event-driven computation and communication result in significant power savings since portions of the hardware modules not receiving fan-in signals can be turned off. The power

3.1 Synapse Design and Synaptic Action

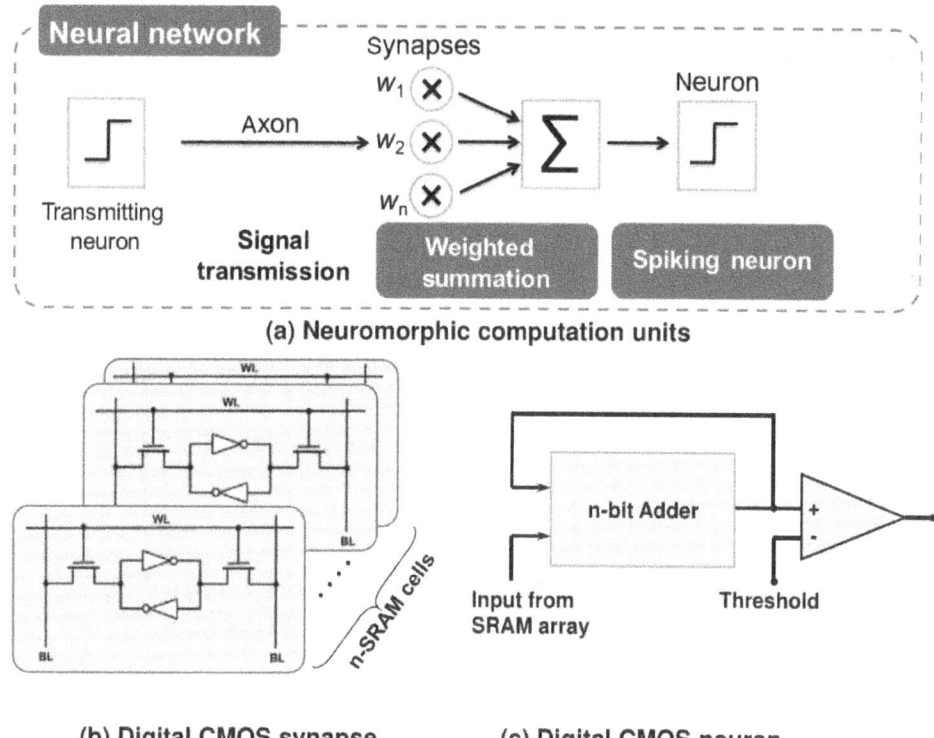

Figure 3.3 (a–c) Digital CMOS implementations of a neuron and a synapse [6]. Adapted with permission from [6]. Copyright © 2017, IEEE.

and energy benefits resulting from such computational modeling effects are purely stemming from the algorithm layer of the design stack and equally applicable for CMOS as well as post-CMOS device technologies.

As we transition next to silicon-based implementation of synapse functionalities, it is vital to highlight some of the basic functions of an artificial synapse:

(i) A synapse needs to modulate the strength of input being propagated from a fan-in neuron to a receiving neuron.
(ii) It needs to continuously update its strength based on the loss function being defined for the particular network architecture. The update can either be local or global depending on the learning algorithm being used (discussed in Chapter 7).
(iii) It needs to retain the updated state till the next programming iteration or over longer durations of time for inference-only hardware engines. In case long-term retention is not possible, state decay over time needs to be considered during learning algorithm formulation (mentioned later in this section).

Next, let us try to understand the challenges stemming from a sole CMOS implementation. Figure 3.3 depicts the hardware implementation of a neuron and a

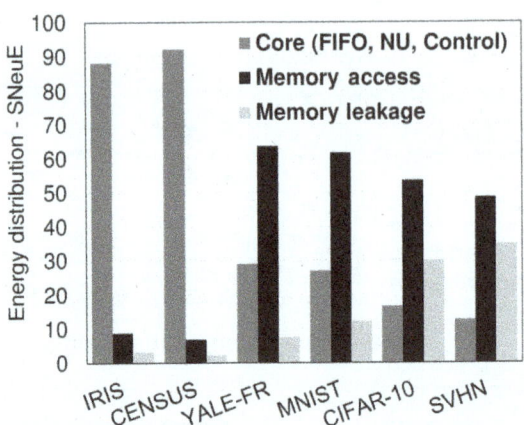

Figure 3.4 Energy consumption profile of a von Neumann architecture-based CMOS machine learning hardware accelerator [6]. Adapted with permission from [6]. Copyright © 2017, IEEE.

synapse unit [5, 6]. As outlined before, implementation of a neuron would require operation of a Multiply and Accumulate Unit which consists of a large number of adders and multipliers. From a synapse perspective, every bit would require a 6T/8T SRAM (Static Random Access Memory – two inverters connected in feedback forming a memory element with peripheral transistors for reading and writing) cell, thereby requiring 24–32 CMOS transistors for a 4-bit synapse (typically, for inference one can reduce the precision to 4 bits without suffering accuracy degradation). The primary reason behind this functional mismatch is that ON–OFF CMOS transistors are essentially suited for implementing Boolean logic type of function implementations and may not inherently map to the neuro-synaptic functional elements required in deep learning applications.

In addition to the abovementioned functional mapping mismatch between the computing kernel and hardware, CMOS systems also suffer from the so-called von Neumann bottleneck. Since deep learning models are memory dominated, a lot of energy is wasted in data transfer between the memory and compute units. Figure 3.4 shows the energy consumption profile for a CMOS hardware accelerator running different pattern recognition workloads where the complexity of the workload increases from the left to the right [6]. As we see from the plot, more than 50% of the total energy consumption can be attributed to just memory access and memory leakage operations for complex machine learning workloads.

In stark contrast, recent advancements in nanoelectronics have revealed several post-CMOS technologies where devices can emulate various neuro-synaptic functionalities through their intrinsic device physics at low terminal voltages.

3.1 Synapse Design and Synaptic Action

These devices are generally referred to as memristive devices [7] since they are usually nonvolatile (and therefore can serve as memory elements) and may be programmed to multiple resistive (or equivalently, other read output) states. Further, when arranged in an array structure as shown in Figure 3.1(b), they can directly implement the dot-product computing kernel required in such applications. For instance, let us assume that each horizontal voltage line represents the corresponding input in a particular layer of the neural network. Next, let us consider that the synaptic weights are programmed as conductance values at each cross-point of the array. Therefore, the current flowing through each cross-point will be the product of the applied voltage and the corresponding conductance. All these currents will get summed up along the column of the array due to Kirchhoff's law, thereby realizing the Multiply and Accumulate functionality in the analog domain. Further, we can circumnavigate energy wastage due to the von Neumann bottleneck since these are nonvolatile devices and therefore enable "In-Memory Computing."

Sections 3.2–3.8 will introduce the material and operational basics of several key device technologies that emulate synaptic characteristics at a single or few devices level. One term that will be used in various sections is "memristive" which refers to the device having a resistance state (or equivalently an order parameter) which is history-dependent and typically nonvolatile. While this terminology was initially introduced in the context of electrical resistors with memory by Chua in the early 1970s, their usage has expanded to other device technologies combining memory and experience. The reduction in number of devices compared to CMOS implementation schemes arise from the rich physics at the material and device levels either intrinsic to the system or due to extrinsic impurities. Where applicable, we also point out how the structural degree of freedom enables new functions that enhance the synaptic characteristics. Here, we outline some of the key performance metrics that can be used to characterize these emerging technologies:

(i) **Number of programmable states**: As noted earlier, one of the key reasons underlying the inefficiency of CMOS systems lies in the binary state representation of CMOS logic. Higher the number of programmable states in a memristive device, higher is the application-level accuracy. However, this usually comes at the cost of device scaling. As devices are scaled down, the bit precision available in memristive devices starts reducing. Application-level requirements also play a key role. Usually, four bits or less are required for inference-only scenarios, while higher bit precision (eight bits or more) are required for on-chip training. Research is ongoing to reduce the bit precision requirement during training [8] as well as inference [9].

(ii) **ON–OFF ratio**: Given weight magnitudes lying between zero and the maximum value to be mapped linearly to the synaptic conductance range, very small weights need to be clipped off to the minimum conductance state of the device. Therefore, higher the ON–OFF ratio of the device, more is the range available for weight programming. Usually, a ratio of 10 or more is required from an application perspective.

(iii) **Device area**: As mentioned earlier, device area is determined usually by the number of states required in the device.

(iv) **Linearity during read**: The basic implementation of dot product in the in-memory computing kernels of memristive crossbar array relies on the notion of Ohm's law being valid. However, many technologies are highly nonlinear and therefore may result in nonideal dot-product calculation. Further, wire parasitics in the array may itself result in some degree of nonlinearity.

(v) **Linearity during write**: The change of device conductance with respect to the amplitude of a programming pulse or duration may be linear or nonlinear. In some cases, such linearity is desired to map software algorithmic modules to the device-circuit operation [10]. In other scenarios, there are opportunities in redesigning the algorithm itself to map to the intrinsic device characteristics (which may include nonlinearities) [11]. For instance, nonlinear synaptic conductance updates can potentially reduce the number of training samples required during the learning process [11].

(vi) **Decoupled read and write paths**: In on-chip learning scenarios, decoupled device read and write paths are usually desired to optimize the two paths separately.

(vii) **Variability**: Variability is usually a concern when designing neuromorphic hardware. Cycle-to-cycle (C2C) variations can be usually leveraged from an algorithm standpoint to design probabilistic computing frameworks [12]. However, developing fabrication process flows to minimize device-to-device (D2D) variations is critical. One approach to deal with D2D variations is through hardware-in-the-loop training [13].

(viii) **Retention**: Retention requirements for the device vary depending on the functionality. Synapses are memory elements and therefore require larger data retention. Depending on the frequency of learning updates, state retention requirements in synapses can vary from milliseconds to days and even longer. On the other hand, neuronal devices are compute elements and therefore can operate much faster at even approximately nanosecond timescale. Device speed is also intrinsically tied to device area. Therefore, neuronal devices can be scaled down further in principle than synaptic devices. Some works have leveraged state decay observed in certain synaptic device technologies for implementing short-term memory effects – useful for lifelong learning [14].

(ix) **Endurance**: Endurance varies a lot across the spectrum among device technologies. For instance, spintronic devices can offer high endurance up to $\sim 10^{12}$–10^{15} cycles while other technologies like phase-change memories have lower endurance levels up to $\sim 10^7$–10^9 cycles.

(x) **Array scaling**: Usually, array size is limited due to parasitic effects as well as device characteristic constraints.

(xi) **Peripheral circuitry**: Each device is interfaced with access transistors or selector devices. Further, the array itself is interfaced with analog-to-digital and digital-to-analog converters since the computation occurs in analog mode inside the array and data communication between arrays needs to take place in a digital fashion. Device-specific operation can drive the design of these peripherals which may, in some cases, end up with more power and energy consumption than the core array itself.

Next, we will discuss synaptic devices and representative characteristics demonstrated to date across a broad spectrum of inorganic and organic materials that may form neuromorphic hardware fabrics. We emphasize the basic principles resulting in synaptic action in each case while providing specific examples for illustrative purposes.

3.2 Filamentary Synapses

A wide variety of materials systems beyond silicon CMOS circuits have been demonstrated for synaptic functions. Creating a three-layer stack that comprises of two metal electrodes separated by a thin (leaky) dielectric represents among the simplest and most widely studied synapse structures. In theory, this is nothing more than a capacitor that has finite leakage and can undergo soft breakdown. However, the internal dynamics of such a device with a carefully prepared dielectric and electrodes chosen for specific reactive properties can enable several synaptic actions. In almost all these devices, the effective resistance of the dielectric layer is tunable by formation of filamentary conducting pathways due to defects. The nature of defects and their formation mechanisms, however, can be drastically different. Much work has been done on binary oxides such as TiO_2, HfO_2, and Al_2O_3 due to the relative ease of thin film materials synthesis and device fabrication [15]. Indeed, for proof-of-concept studies and understanding the operational principles, such devices can be made simply with shadow masks without the need for complex lithography. Further, the vacuum deposition tools needed for making a two-terminal leaky capacitor is also modest and much of the fabrication process can be completed without access to sophisticated cleanrooms. This has in part led to an explosion of research in synapses and several dozen (if not more) of material classes can be utilized to realize different forms of synaptic plasticity. In this

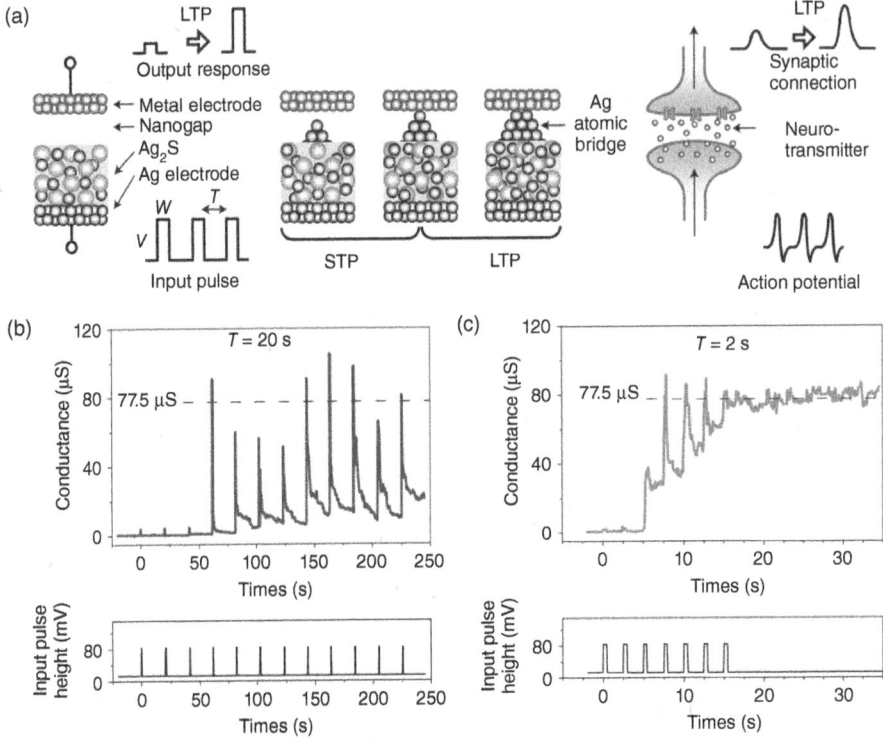

Figure 3.5 (a) Schematic of Ag$_2$S synapse and the biological counterpart, (b) change in conductance of the synapse when an input pulse of 80 mV, width of 0.5 s is applied every 20 s, and (c) identical stimulus applied every 2 s. In the former case, short-term plasticity is noted while in the latter, long-term plasticity is noted. The increase in conductance due to single atom contact is $2e^2/h$ corresponding to 77.5 μS [16]. Adapted with permission from [16]. Copyright © 2011, Springer Nature Limited.

chapter, we describe a few experimentally reported artificial synapses focusing on the overarching mechanisms. This is by no means a complete list. Review papers that compare various synaptic devices and the forms of plasticity are listed at the end of the chapter for the interested reader. First, we consider a Ag-based atomic switch that can form across two electrodes leading to formation of conducting filaments [16]. Figure 3.5 shows a two-terminal device comprising of Ag electrode with a Ag$_2$S layer grown on top.

The top electrode is Pt and formed with a scanning tunneling microscope tip. A small gap exists across the Ag$_2$S and the top electrode forming a nanogap. By application of a bias, it is possible to precipitate Ag at the surface of the Ag$_2$S forming a conducting bridge. Once the bridge is formed, a current path is created leading to modification of the initial resistance state. Further, by adjusting the strength and periodicity of the electrical stimulus, it was found that both short-term and long-term plasticity could be realized. Weaker electrical contact enables spontaneous forgetting (decay of resistance state) due to break in the atomic contact and opens opportunities to emulate human memory recall-related research. This is due to the controllable stability of the conducting bridge that was formed. Hence, a simple two-terminal device could mimic fundamental aspects of synaptic learning noted in neuroscience studies.

Oxygen vacancy migration across electrodes in a dielectric layer represents another popular class of devices that have been examined for artificial synapses. The dielectric layer can be a binary oxide, such as hafnia, titania, tungsten trioxide, or ternary oxides such as $SrTiO_3$ and prepared in an oxygen-deficient composition

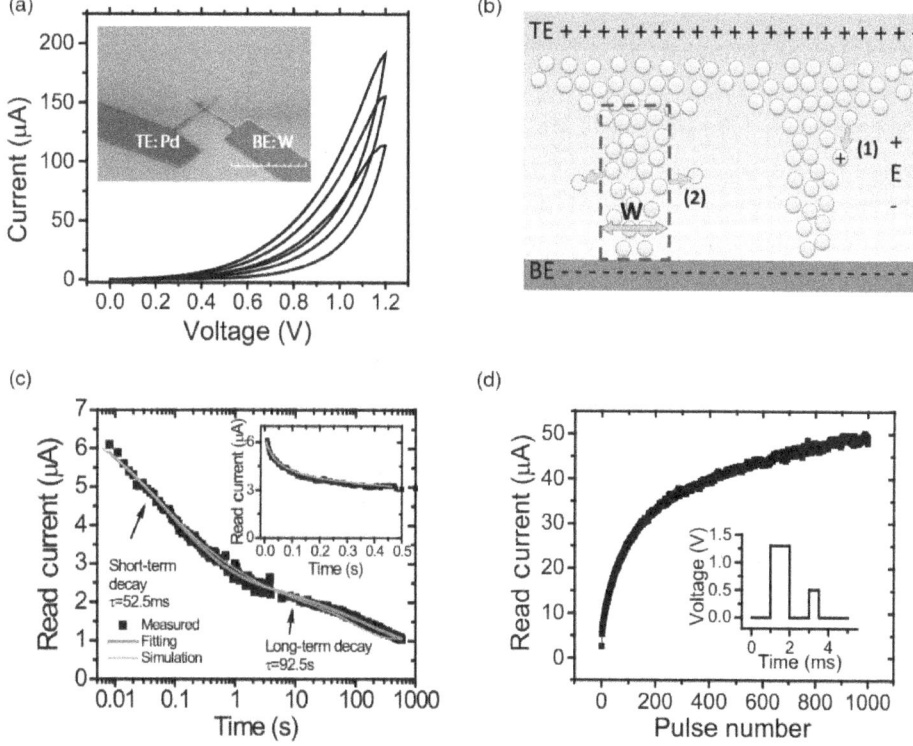

Figure 3.6 (a) Inset shows schematic of a two-terminal device with W as bottom electrode and Pd top electrode. The switching layer is WO_3. The DC I–V characteristics show pinched loop and continuous tuning of the resistance over multiple cycles due to the drift of the oxygen vacancies. (b) Schematic showing drift of oxygen vacancies as well as spontaneous diffusion, (c) conductance decay of the memristor device after being subject to 10 pulses of 1.2 V, 1 ms duration, and (d) continuous tuning of resistance by applying successive electrical pulses of 1.3 V, 1 ms duration and the resistance is readout using 0.5 V, 500 μs pulse. The variation in read current with time can be modeled as a stretched exponential and fit to device models [17]. Adapted with permission from [17]. © 2015 WILEY-VCH Verlag GmbH & Co. KGaA, Weinheim.

such that oxygen vacancies are present in the initial as-grown film. The oxygen vacancies are charged (in Kroger–Vink notation, they are written as $V_{O^{..}}$ where the two dots in superscript represent two positive charges) and therefore drift under electric field applied across the electrodes. Upon formation of a conduction pathway at a critical defect density, the device undergoes a soft breakdown and results in a sharp change in conductance. The change in conductance can be long-lasting (i.e., nonvolatile) or short-term resulting in a decaying memory. By application of opposite polarity voltage, it is then possible to reset the device into the original high-resistance state (HRS). Further, by applying incremental electrical pulses, it is possible to gradually modify the electrical resistance that is useful to demonstrate potentiation–depression type tuning of resistance. In Figure 3.6, one representative example of weight modulation in WO_3 layers due to motion of oxygen vacancies is presented [17].

Regions rich in oxygen vacancies form high-conductance channels while those with lesser defect density form high-resistance paths due to Schottky contacts. The parallel paths for current transport can be modeled appropriately and the internal dynamics offer multiple applications as synapses. The gradual decay of resistance when the stimulus is removed offers programming versatility. The drop in resistance with time has both short-term and long-term time constants from about 52 ms to 92.5 s providing an analogy to short-term and long-term plasticity in biological synapses. Further the change in resistance is nonlinear and saturating likely due to consumption of oxygen vacancies. Such behavior can be further utilized to demonstrate paired-pulse facilitation (PPF) and spike-timing-dependent plasticity.

It is important to note that the electrodes also play an important role in determining the operation of synaptic devices that utilize filamentary paths. This can occur through several mechanisms. First, the electrode(s) could react with the dielectric layer and scavenge oxygen out of the dielectric creating a local oxygen-deficient layer at the electrode–dielectric interface region that can nucleate switching filaments. Materials such as Ti and Al that have high thermodynamic propensity for oxidation are well suited for this purpose. The electrode can offer a Schottky barrier selectively depending on the stoichiometry of the dielectric and/or work function tuning and therefore affect the conduction pathways. In this case, the low-resistance state (LRS) is enabled by tunneling across the interface, while HRS is due to the finite width of the depletion layer when the voltage polarity is reversed [18, 19]. Finally, the electrode itself could dissolve into the dielectric at an atomic scale resulting in metallic filaments that percolate across the dielectric layer.

The insulating medium could further be replaced with an electronic composite comprising dielectric–metallic inclusions that are non-percolating to begin with. Figure 3.7 shows current–voltage characteristics of a Pt-oxide:Ag-Pt sandwich structure [20]. Ag nanocrystals are embedded in the oxide dielectric that could be any of MgO, SiO_xN_y, or HfO_x. In contrast to nonvolatile bipolar memory

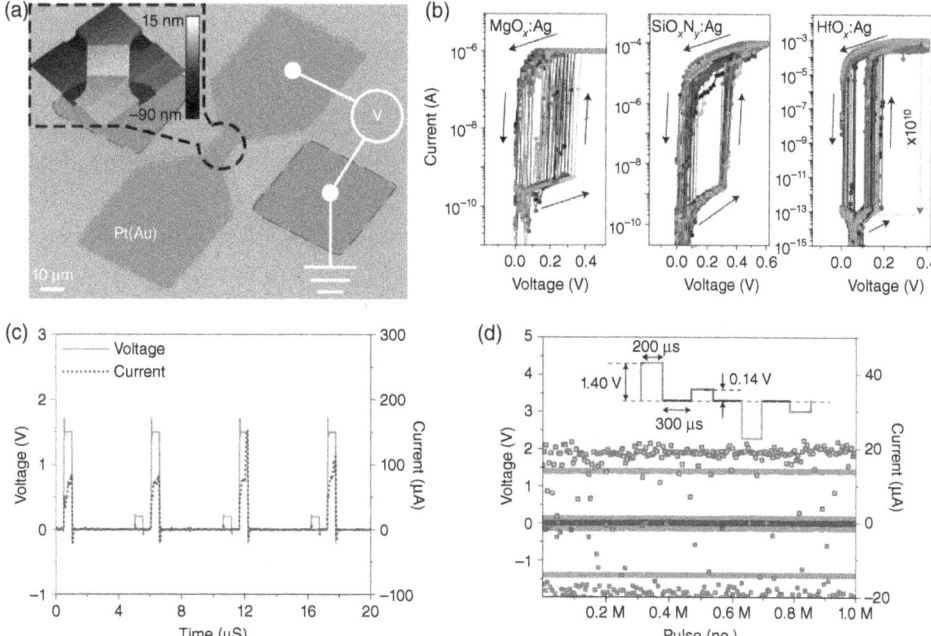

Figure 3.7 (a) Scanning electron micrograph of a two-terminal device with inert electrodes such as Pt or Au. Top electrodes are positioned such that they go top right to bottom left in the figure, and bottom electrodes are positioned such that they go from top left to bottom right, respectively. The inset shows an atomic force microscope image of the junction. (b) Shows threshold switching of channel resistance at critical voltage for different dielectric media embedded with Ag nanoparticles. (c) Shows the electrical switching characteristics measured at 600 K. The device relaxes back to the insulating state within 5 μs when the voltage pulse is removed. (d) Switching endurance for 1 million cycles at room temperature for positive and negative voltage polarities [20]. Adapted with permission from [20]. Copyright © 2016, Springer Nature Limited.

switches that frequently utilizes oxygen vacancy migration, the sharp rise in current decays spontaneously when the voltage is removed in all cases.

The electrical characteristics are unipolar in nature from observation of symmetric hysteresis loops. The resistance ratio between the insulating and conducting states can be impressively high, ranging from 5 to 10 orders of magnitude while the turn-on slope can be as steep as 1–10 mV/decade. Electron microscopy of planar junctions to understand the mechanisms governing the dynamics reveal an interesting property, namely, the growth of Ag conducting bridges leads to the abrupt increase in current resulting in the threshold switching. At the same time, when the voltage is removed, the silver nanoparticles spontaneously contract forming clusters to minimize the interfacial energy between the metal and the dielectric media. The application of the electric stimulus serves as a mechanism for Joule heating of the channel region providing energy for the nanoparticle clusters to break up and distribute across the electrode gaps forming LRS. Such a mechanism offers several interesting directions to demonstrate synaptic action. Sequential application of electric stimulus can gradually alter the location and shape of the conducting bridges. Maintaining a short-term memory due to thermal dissipation timescales leads to pulse-dependent and time-interval-dependent plasticity of the device resistance. Hence, a composite medium where interfacial energy and wetting of nanoparticles in a matrix can be carefully controlled offers route to synaptic memory for further exploration in neural biomimetics. The dynamics of such devices, particularly the resetting process to initial state will be strongly dependent on temperature; therefore this presents a limitation if large thermal loads are expected.

3.3 Ferroelectric Synapses

Ferroelectric (FE) materials are insulators possessing spontaneous polarization that can be switched by electric fields. The polarization arises typically from lack of centrosymmetry and leads to "polar" crystals that may potentially host FE properties. The ability to encode information based on the orientation of the polarization vector has been historically utilized in FE memory device technologies. Much work has been carried out in both academic and industrial settings on nonvolatile memory with ferroelectrics over the past several decades. Well-separated binary resistance or capacitance states (corresponding to assigned UP and DOWN polarization vectors with respect to an electric field) are useful for design of nonvolatile memory. The two orientations of the polarization can correspond to "0" and "1" binary states. In the context of neuromorphic electronics, having an additional degree of freedom to realize continuous tuning of resistance states is desirable for synapse design. In this section, we discuss a few representative examples of artificial synapses built with FE layers as an active medium. Note that while there are numerous materials that possess FE properties originating from crystal symmetry considerations, in practice only a few turn out to be technologically relevant. This down selection arises due to combination of several factors including ease of monolithic integration on silicon platforms for on-chip or backend of line (BEOL) application purposes, need for high Curie temperature (temperature above which the spontaneous polarization disappears and the material transitions into a paraelectric), chemical stability of both the material and its interfaces in air or humid atmosphere during processing, magnitude of the remanent polarization, and coercive field that is required to switch the direction of the polarization. Materials that have been studied extensively include $BaTiO_3$ and HfO_2 in pristine and doped form in the realm of inorganic materials and PVDF as a polymer material. We discuss inorganics first followed by organic FE synapses.

A tunnel barrier comprising of an FE layer sandwiched between two conducting electrodes is one approach to design synaptic devices. These are also referred to as ferroelectric tunnel junctions (FTJs) and were reported as early as 1971 by Esaki and coworkers. When the polarization is switched by an electric field, the resistance for electrons to tunnel across the barrier is different, resulting in tunnel electroresistance effect. Asymmetric potential barriers at the FE interfaces result in differing potential barriers and hence result in polarization-dependent junction resistance. While such an effect can be readily envisioned for binary data storage, multilevel and continuous tuning of resistance is essential for synaptic behavior. For this purpose, it is necessary for the FE domain microstructure to be designed carefully.

Particularly, the domain configurations in the FE must be crafted such that variable fraction of the oriented domains are possible under sequence of electric stimuli. Figure 3.8(a) shows resistance plotted as function of voltage taken from 2 nm thick $BaTiO_3$/30 nm $La_{0.67}Sr_{0.33}MnO_3$ bilayer film grown on insulating $NdGaO_3$ substrate [21]. The top electrode was Co/Au film of 350 nm diameter patterned by electron-beam lithography. The pulse duration was fixed at 20 ns while the amplitude of the pulse was systematically varied. The ON/OFF ratio approaches 300 when the

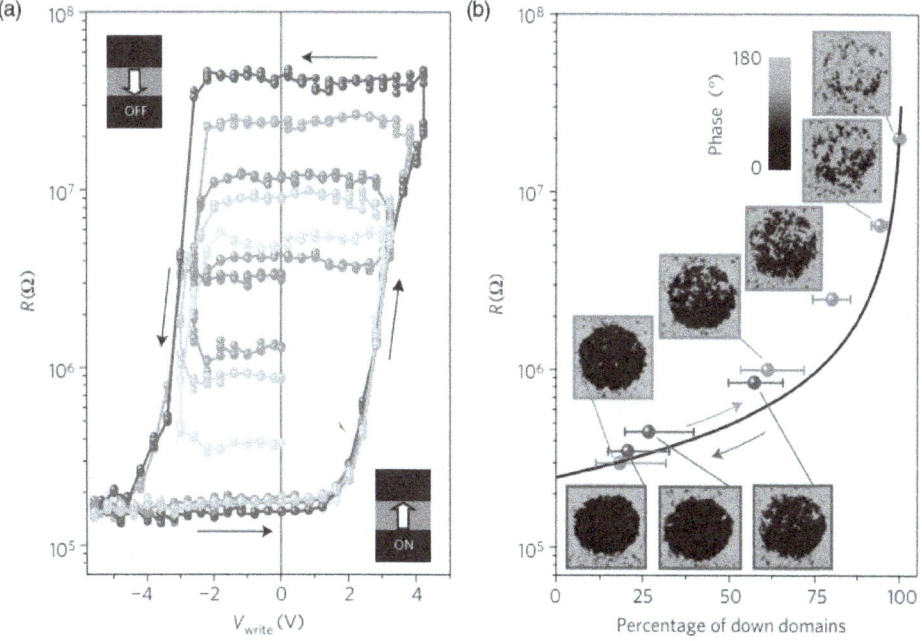

Figure 3.8 (a) Junction resistance measured after application of 20 ns voltage pulses of different amplitudes. The read voltage is 100 mV. The curve is asymmetric with respect to the zero of the voltage bias due to asymmetry in the tunnel barrier at either interface with the ferroelectric. (b) Change in junction resistance plotted as function of percentage of down domains. The domain fraction is determined from the PFM phase images. The ability to vary the relative fraction of up or down domains enables the analog tuning of the resistance [21]. Adapted with permission from [21]. Copyright © 2012, Springer Nature Limited.

voltage is swept between −5.6 V and +4.2 V (the outermost curve). Interestingly, intermediate resistance states are observed as the voltage is modulated through smaller ranges building up hysteretic curves and minor loops. The origin of minor loops can be understood from Figure 3.8(b) that shows piezoresponse force microscopy images from the sample taken at different resistance states. Starting from LRS corresponding to up-polarized domains, down-oriented domains begin to nucleate under positive voltage bias stimulus. The domains expand under consequent bias and contribute to the resistance evolution. The black curve represents a parallel conduction model considering up and down polarized domain fractions. Such devices also exhibit pinched hysteresis loops displaying memristive properties.

Figure 3.9 shows the continuous control over the resistance by adjusting the voltage pulse profile [21]. As described in the caption, the number of voltage pulses corresponding to positive or negative polarity can modulate the resistance. Hence, amplitude of the voltage pulse and number of pulses can both affect the

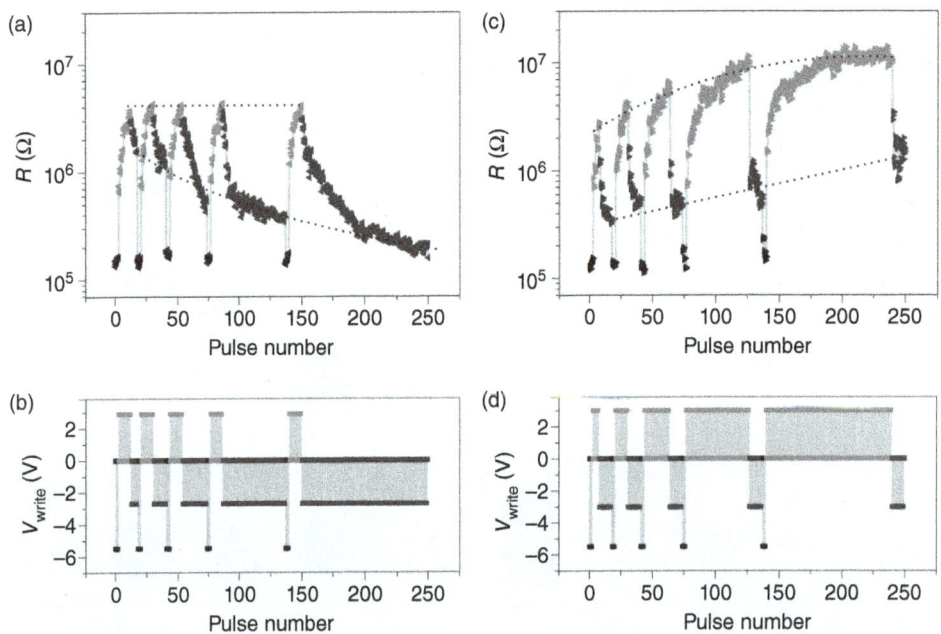

Figure 3.9 The junction resistance can be continuously tuned by application of pulsed voltage stimuli. (a) Resistance of tunnel junction reaches identical value when the number of positive voltage pulses is fixed at 10, while the final state resistance depends on number of negative pulses. (b) Shows the voltage pulse scheme for positive and negative biases for data presented in (a). (c) Variation in resistance as the number of positive pulses is increased. The resistance under negative bias depends on the starting point at the high-resistance state. (d) Shows the voltage pulse profile used to measure resistance evolution in (c) [21].
Adapted with permission from [21]. Copyright © 2012, Springer Nature Limited.

resistance of the junction and forms one basis to realize spike-timing-dependent plasticity. Hence, fine control over fraction of domains oriented along a direction is central to synaptic properties of FTJs. At the same time, understanding factors that control the junction resistance for different bias polarity ultimately sets the limit for number of states that might be measurable or useful in a network.

Since the report of ferroelectricity in hafnia about a decade back, interest in examining their integration into CMOS technologies has grown significantly. This is because hafnia has been extensively researched in the past two decades as a replacement gate dielectric for SiO_2 and monolithic integration onto Si CMOS using atomic layer deposition methods is now somewhat mature. While much work was conducted on amorphous or partially crystalline hafnia to examine the dielectric constant and leakage characteristics, ferroelectricity has been observed in hafnia crystals with metastable phases such as orthorhombic structures [22]. Note that the typical crystal structures found in pure hafnia films are monoclinic or tetragonal depending on grain size and film thickness. Therefore, often the hafnia film is alloyed with Zr, Si, and so on to enable the formation of metastable structures. Hafnia as an FE is therefore emerging as an interesting material to explore in the context of neuromorphic computing particularly with an eye to back-end-of-line compatibility and material familiarity in the semiconductor industry. Tunnel junctions with $Hf_{1-x}Zr_xO_2$ (HZO) have been studied as artificial synapses. To avoid leakage currents that might overwhelm the polarization dependence of tunneling current, researchers have developed the concept of dual layers comprising an FE and dielectric stack sandwiched between electrodes. An example is introducing a thin alumina dielectric grown on an HZO film. With this structure, a significant fraction of leakage current is blocked while the effect of switching the domain configuration in HZO can still affect the tunnel junction resistance resulting in nearly 10× change in resistance [23].

In addition to orthorhombic structures, there are other possible metastable phases that can emerge due to doping or strain. HZO film grown on epitaxial $La_{0.7}Sr_{0.3}MnO_3$ (LSMO) films grown on (001) $SrTiO_3$ substrates has a rhombohedral structure due to compressive strain imposed by the substrate. Using LSMO also as a conducting top electrode, synaptic weight modulation has been studied in this system. Figure 3.10 shows different potentiation–depression profiles that can be obtained with the same sample by varying the voltage pulse–time width profiles [24].

Figure 3.10(a) and (b) shows the device structure and electrical test profile used for the weight update measurements. The experimental protocol, referred to as PUND (positive-up-negative-down), is used to obtain the intrinsic FE switching. The first positive pulse switches the polarization to one direction, while the second positive pulse enables eliminating spurious contribution from leakage currents and charging artifacts. The third pulse, which is in the opposite direction,

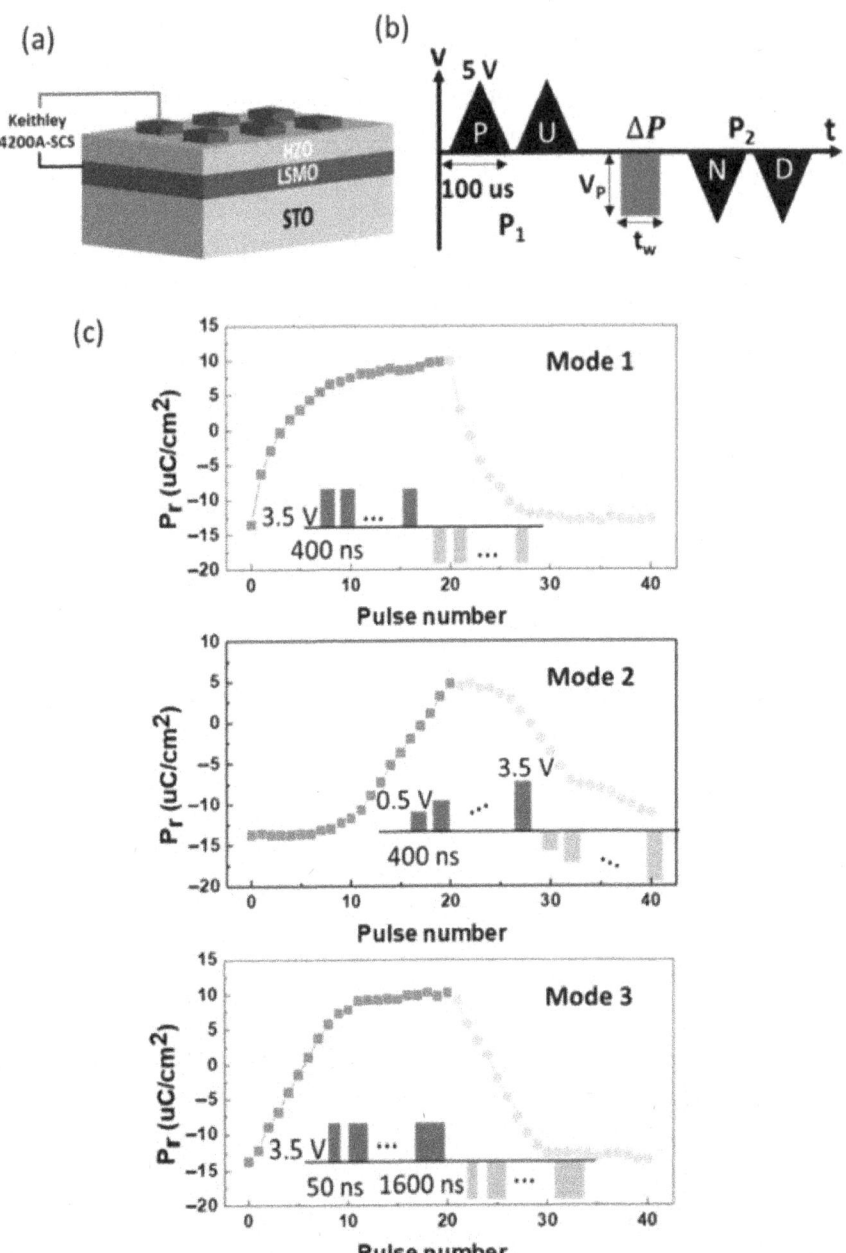

Figure 3.10 (a) Schematic of the two-terminal device structure used for synaptic measurements and (b) programmed voltage–time profile used for the weight measurements. The effective change in polarization due to the domain switching can be obtained from this experiment. (c) Remanent polarization versus pulse number for different measurement modes. The modes 1–3 correspond to distinct voltage–time profiles as described in the text. Note that the shape of the potentiation–depression curves and saturation are quite different in each case [24]. Adapted from [24], under the Creative Commons Attribution 4.0 License. CC BY 4.0.

switches the polarization direction for some fraction of the domains, and the fourth pulse enables measurement of the polarization that was left to be switched. Figure 3.10(c) shows variation in the remanent polarization as function of testing mode. Mode 1 corresponds to identical pulse train of 3.5 V amplitude with time interval of 400 ns, while Mode 2 corresponds to increasing pulse amplitude from 0.5 V to 3.5 V while keeping the interval constant at 400 ns. Mode 3 corresponds to same amplitude of 3.5 V and increasing time interval from 50 to 1,600 ns. Clearly the profile and symmetry of the remanent polarization curves can be controlled by the electrical testing protocols. Linearity, number of states before saturation are all different in each case. By combining two voltage spikes applied to the top and bottom electrodes representing presynaptic and postsynaptic events, it is possible to emulate spike-timing-dependent plasticity. While each voltage can be smaller than the coercive voltage, if the profiles overlap in time such that the combined magnitude is large enough to switch sufficient population of domains, then the remanent polarization will show a clear dependence on the time interval between pulses emulating learning. The plasticity due to electric stimulus programming arises from the kinetics of domain nucleation leading to polarization switching and is related to the interaction of point defects such as oxygen vacancies with the electrodes. By choosing crystal orientation of the top and bottom electrodes, it is possible to tune the thermodynamics and kinetics of charged defect migration across them, thereby affecting the kinetics of local electric field distribution and polarization dynamics. Hence, not only is the FE layer important, but also carefully selecting electrode materials for maximizing the ON/OFF ratios and intermediate state dynamics. Ferroelectric field-effect transistors (FeFETs) comprising the FE layer as a gate stack component in series with a resistor also are well suited to design synapses [25]. In such architectures, the polarization of the FE can be modulated by external field, and the corresponding gate-drain resistance and the threshold voltage for inversion of the adjacent semiconductor channel can be controlled in an analog manner. Successful demonstration of FeFET-based synapses with silicon-doped hafnia has been reported in literature indicating promise for further development. The critical microstructural features that enable synaptic action include control over the domain orientation and polarization control that are necessary to realize continuous tuning of the threshold voltage.

Organic FE polymers are also being actively studied for neuromorphic applications. A popular material system extensively investigated for this purpose is P(VDF-TrFE) and is discussed as an example. Unique advantages of this material include permanent dipole due to the presence of electronegative fluorine and electropositive hydrogen in the polymeric chain and scalability down to few layer thicknesses while retaining ferroelectricity. This is an advantage over inorganic systems where FE properties severely degrade when the thickness is reduced to nanometer length scales. Hence, for devices such as tunnel junctions, it is quite challenging

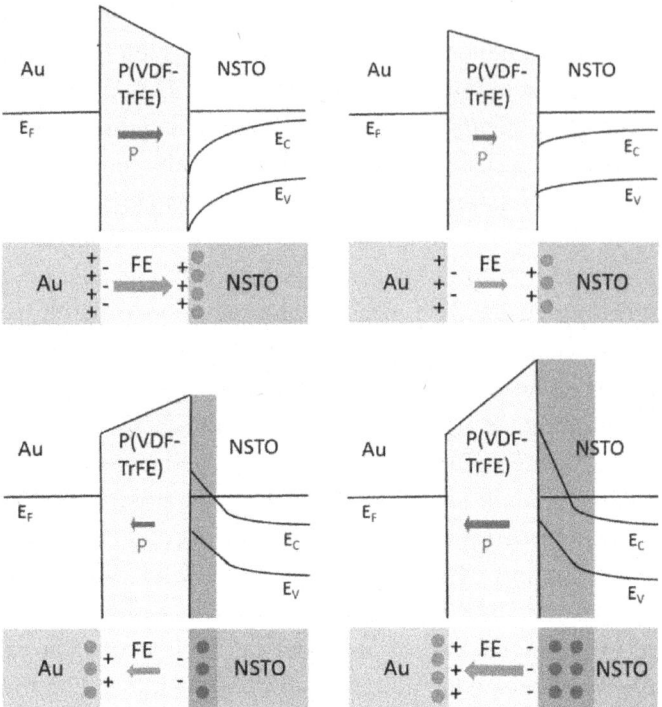

Figure 3.11 Band diagram of Au-P(VDF-TrFE)-Nb:SrTiO$_3$ heterostructure. The barrier heights for electron transport across the interfaces depends on the polarization vector orientation. In one case, there is a Schottky barrier corresponding to depletion of majority charges at Nb:STO interface, and this is eliminated under positive bias to the substrate [26]. Adapted from [26], under the CreativeCommons Attribution-NonCommercial-NoDerivs License.

to work with oxide ferroelectrics unless special single crystal substrates are used to stabilize the phase. Polymers also have the advantage of fabrication using liquid precursors, hence eliminating the need for expensive ultrahigh vacuum equipment. Again, to expand the properties from bistable resistance switching to multistate analog mode, it is important to control the domain fraction in the two polarization states. Figure 3.11 shows the band diagram for an Au-P(VDF-TrFE)-Nb:SrTiO$_3$ heterostructure [26]. When the polarization vector points toward the Nb:STO, electrons accumulate at the interface. When the polarization vector points away from the Nb:STO, electrons are depleted at the interface resulting in formation of a Schottky barrier. Hence, the two polarization orientations result in different transport characteristics for charge carriers, resulting in distinct resistance states. This explains the bistable resistance switching and as expected, if the relative fractions of the polarized domains can be varied, the effective resistance can be modulated between the two extreme values for complete polarization rotation.

3.3 Ferroelectric Synapses

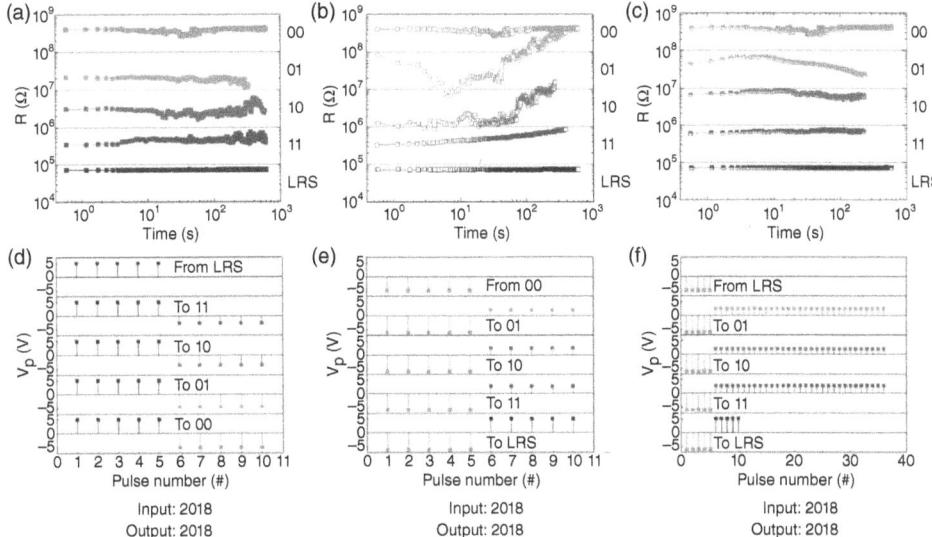

Figure 3.12 (a–c) Intermediate resistance states being accessed by voltage pulse trains shown in (d–f), respectively. In each case, the stability of the state is monitored over time. An asymmetry in resistance stability is seen when transitioning from LRS–HRS and vice versa. Also, long pulse trains induce greater stability in resistance state. These influence the data storage and retrieval process. Fluctuations in intermediate states can corrupt the stored data resulting in errors during retrieval. This is seen in (b) where the input year is 2018, whereas it is mistakenly retrieved as 1004 [26]. Adapted from [26], under the CreativeCommons Attribution-NonCommercial-NoDerivs License.

One example of accessing intermediate states, their storage and retrieval is presented in Figure 3.12. In Figure 3.12(a), the junction is set initially to LRS followed by application of five identical positive voltage pulses (pulse profiles for the different cases are shown in Figure 3.12(d–f)). These pulse schemes put the device in different intermediate states referred to 00 (initial), 01, 10, and 11 where each of these states corresponds to device resistance staying within an order of magnitude. In Figure 3.12(b), the device is set to HRS followed by pulse scheme to reach intermediate states. The fluctuations in the resistance state indicate the instability and corruption of the assigned state. If different pulse profiles are chosen as in Figure 3.12(c), the states are most stable. Simple encoding of numerical values such as the year 2018 in eight-bit ASCII code and their retrieval provides an idea of the stability of the intermediate states. Clearly, accurate retrieval is possible for wisely chosen pulse profiles. This is due to the asymmetry in polarization reversal, the high-to-low resistance transition may be more abrupt than the other way around. Hence, longer pulse trains offer greater chance of transitioning from short-term to long-term plasticity offering more stability and robust storage of intermediate polarization states. Besides the processing ability, it is important to note that polymers that show permanent dipole moment due to the chain elements and do not require ion migration or charge transport can show switching speeds approaching tens of nanoseconds. The intrinsic timescales of domain switching are primarily relevant here since the carrier dynamics at the electrodes could be typically faster and do not limit device performance. Energy consumption for potentiation and depression can approach the 200 fJ range favorably compared to energy consumption in biological synapses.

3.4 Insulator–Metal Transition-Based Synapses

To induce a nonvolatile phase transition, it is often essential to introduce charged defects into the lattice and control their concentration and/or distribution by electric field. The pristine oxide compounds undergo threshold switching and are also often nonpolar. Introducing defects such as oxygen vacancies, protons, and lithium ions allow several interesting properties to emerge. For instance, the conductivity can be gradually modulated by varying the electric field strength. Multilevel states are then possible by introducing different concentrations of defects. The memory state is retained, that is, in a nonvolatile manner. Hence, several distinct properties can be realized when the pristine compound is modified with dopants. In this section, we briefly discuss representative examples where in sequential conductance transition is induced by adding dopants to an oxide channel.

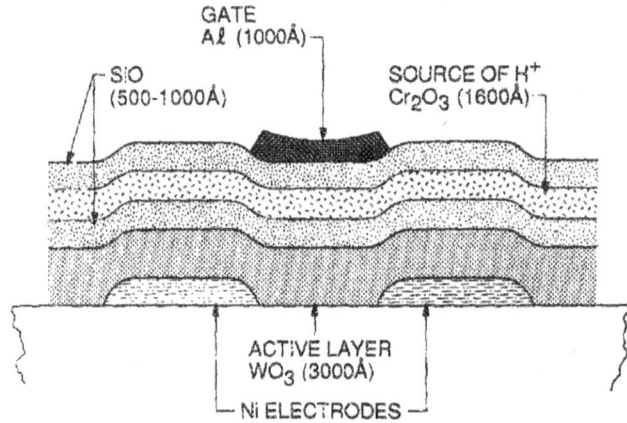

Figure 3.13 Schematic of a three-terminal WO_3 device that functions as an analog synapse. The resistance is modulated in a nonvolatile fashion by introduction of protons into and out of the lattice [27]. Adapted with permission from [27]. 1990 © AIP Publishing.

3.4 Insulator–Metal Transition-Based Synapses

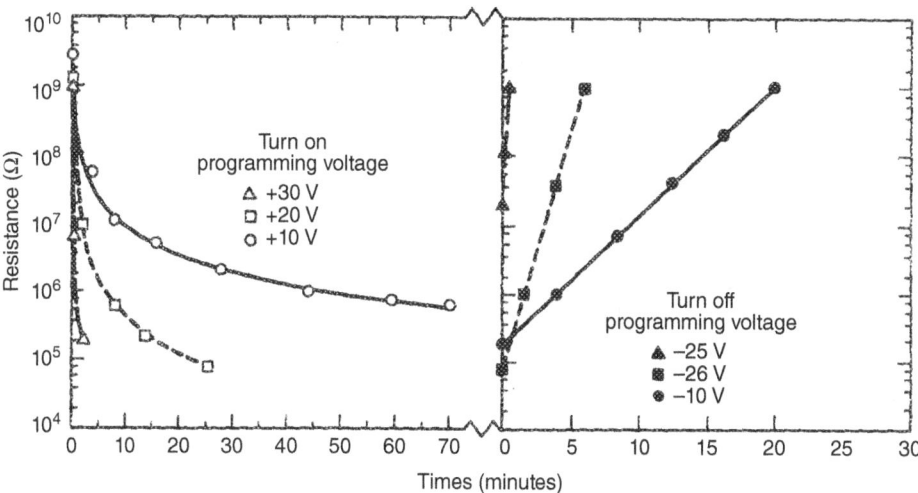

Figure 3.14 Programming characteristics of a device. The turn-on and turn-off characteristics follow different relationships with respect to the driving voltage. A clear dependence of the change in resistance with strength of the electric field is seen in both cases [27]. Adapted with permission from [27]. 1990 © AIP Publishing.

A material that has been well studied for resistance switching by adding dopants is WO_3, a popular electrochromic material. Figure 3.13 shows a schematic of a three-terminal device incorporating tungsten oxide as the channel contacted with two Ni electrodes that act as source and drain [27]. A chromium oxide layer is deposited on top of an SiO layer that is first grown on the WO_3. Another SiO layer is deposited on top followed by an Al gate [27]. Each layer serves certain important roles in operation of this device as an analog synapse. The SiO layers allows proton diffusion while minimizing electron leakage currents. The Cr_2O_3 layer acts as a hygroscopic layer that diffuses protons into and out of the WO_3 channel. The Al acts as a gate electrode. The synaptic function is accomplished by applying various electric field strengths that move protons into and out of the WO_3 lattice. That is, a positive gate voltage drives protons into the WO_3 layer forming tungstic acid (H_xWO_3, where x is atomic fraction), while the reverse polarity moves protons back into the chromium oxide layer. Since the WO_3 layer becomes metallic upon proton injection, a conductance change is realized. This process is both nonvolatile and incremental depending on the electric field strength as shown in Figure 3.14. Some important points to note in this figure are the following: The resistance drop during device turn-on and the resistance increase during turn-off are not symmetric. In fact, power-law behavior was observed for the turn-on while exponential increase in resistance was noted for the reverse process. This asymmetry may be due to the microscopic factors affecting the diffusion of protons into and out of the

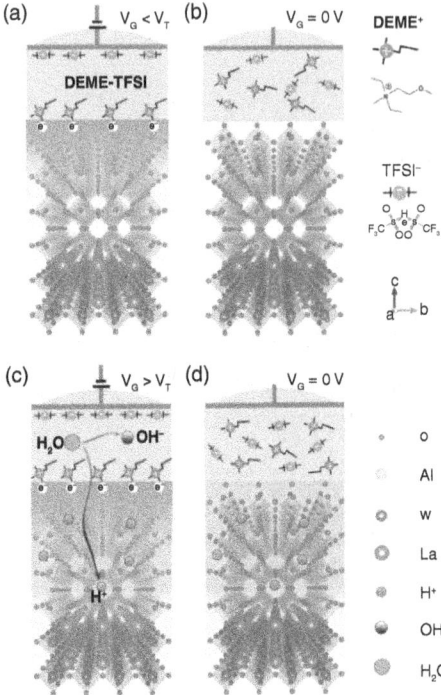

Figure 3.15 (a–d) Effect of applying different electric fields on the carrier accumulation and depletion effect at the IL-WO$_3$ interface. When the voltage is less than a critical value, electrostatic effect dominates resulting in a modest volatile change in resistance. When the voltage exceeds a critical value resulting in hydrolysis of water to produce protons, the protons can then diffuse into the film resulting in nonvolatile modulation of the channel resistance [28]. Adapted with permission from [28]. © 2018 WILEY-VCH Verlag GmbH & Co. KGaA, Weinheim.

lattice. In this initial demonstration, the authors found that the Al gate lasted only around 25 cycles and would blister off. Presumably uncontrolled hydrogen gas evolution out of the device was responsible for this mode of failure.

Besides solid-state dielectrics that act as ion reservoirs, liquids or gels can be applied on top of the oxide surface to modify their electrical properties under electric fields. Figure 3.15 shows schematic of a WO$_3$ epitaxial film grown on LaAlO$_3$ by pulsed laser deposition with DEME-TFSI ionic liquid on top that acts as an electrical insulator and ion conductor [28]. Two regimes of operation can be identified in this device. When the gate voltage V_G is below a critical value (V_T), there is only electrostatic accumulation of charge carriers at the ionic liquid–oxide channel interface resulting in a volatile change in channel resistance. However, beyond a critical value of the gate voltage, water present in the ionic liquid can hydrolyze resulting in formation of protons. The protons can diffuse under electric field into the oxide channel resulting in a large change in channel resistance. This resistance change is

3.4 Insulator–Metal Transition-Based Synapses

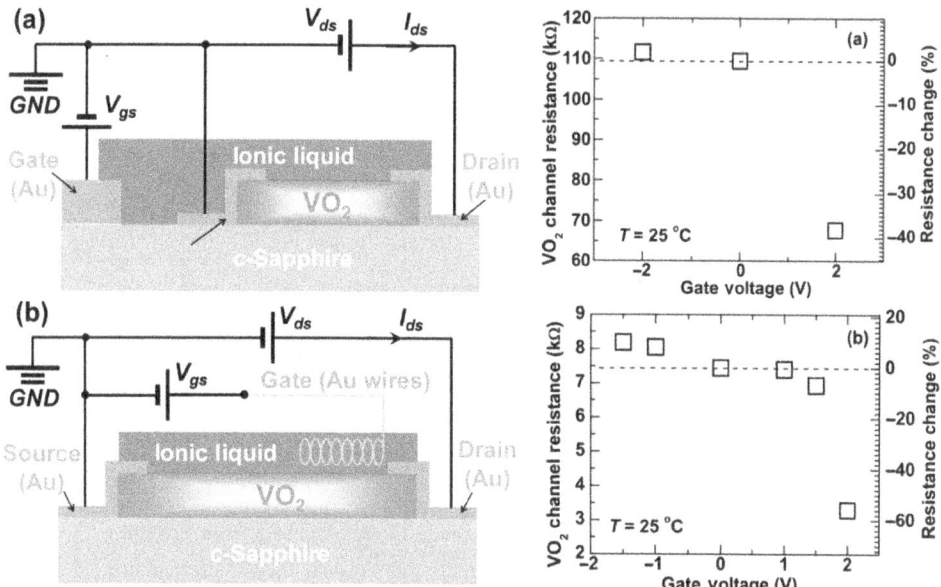

Figure 3.16 Schematic of ionic liquid-gated VO$_2$ devices in (a) planar and (b) vertical geometry. The ionic liquid is used as a gate insulator to apply electric field with the top Au electrode. The figure on the right shows electrostatic modulation of channel resistance as a function of gate bias. The top figure corresponds to channel dimension of 5 mm × 1 mm, while the bottom figure corresponds to that of 40 μm × 160 μm channel. The extent of resistance modulation increases as the channel is scaled [29]. Adapted with permission from [29]. 2012 © AIP Publishing.

also nonvolatile since the presence of the protons ensures this. Only when a reverse polarity voltage is applied to migrate the protons back into the ionic liquid does the resistance regain original value. Hence, similar to a solid gate stack, the ionic liquid offers a simple test platform to probe the effect of charged dopants in influencing the synaptic analog resistance behavior. Using these devices, synaptic functions such as potentiation, depression, short- and long-term plasticity have been demonstrated [28].

Another system that undergoes insulator–metal transition upon ion doping is VO$_2$. We will discuss utilizing the electronic phase transition in the pure stoichiometric compound as an artificial neuron. Now, by adding charged dopants such as hydrogen ion (protons) into the lattice along with electrons for neutrality, we can enable synaptic function. Ionic liquid-gated devices were demonstrated to show electrostatic modulation as well as long-term relaxation (slow dynamics) in VO$_2$ films grown on sapphire. Figure 3.16 shows representative device layout to investigate the charge modulation in electric double layer transistors [29]. When the voltage is low enough to prevent chemical mixing of the ionic liquid–oxide interface, it is possible to observe the expected volatile shift in resistance due to carrier accumulation at the oxide channel surface. As the voltage bias is increased, chemical

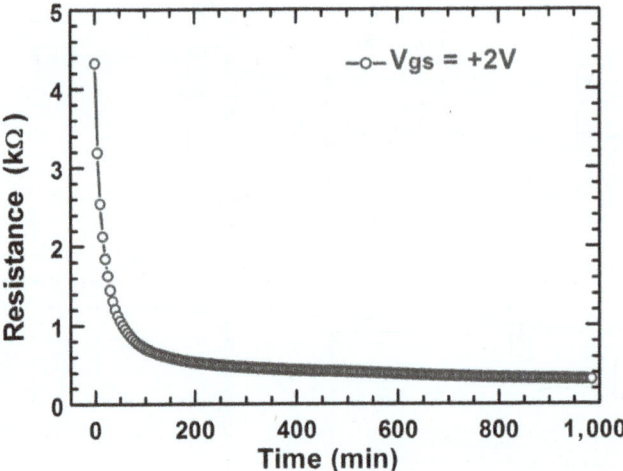

Figure 3.17 Slow relaxation of channel resistance in a VO$_2$ ionic liquid-gated transistor [30]. Adapted with permission from [30]. 2012 © AIP Publishing.

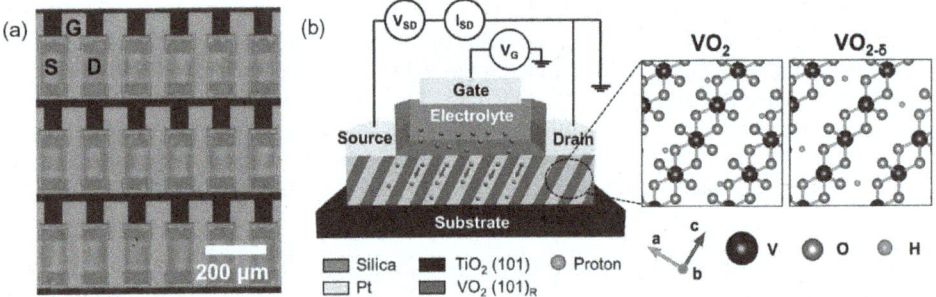

Figure 3.18 Solid-state three-terminal transistor for ion doping. (a) Optical microscope image of devices and (b) schematic of transistor device with electrical connections. The zoomed-in figures show two different cases of channel material, pristine stoichiometric oxide versus an oxygen-deficient channel. The text describes differences in function as the stoichiometry is varied [31]. Adapted with permission from [31]. © 2020 Wiley-VCH GmbH.

instability at the oxide channel interface sets in resulting in mass exchange between the ionic liquid and oxide. This can result in doping the oxide channel with light ions such as protons or oxygen vacancies. As the ions are charged and can act as electron donors, the resistance of the channel is significantly altered due to the drift diffusion of the charges. Hence, there is a non-electrostatic effect leading to time-dependent resistance, which can in principle serve as a synapse. The effect can be utilized to design a volatile or nonvolatile synapse in fact. Figure 3.17 shows the evolution of channel resistance under gate bias [30]. A slow change in resistance with time is observed indicating diffusion dynamics of oxygen vacancies.

3.4 Insulator–Metal Transition-Based Synapses

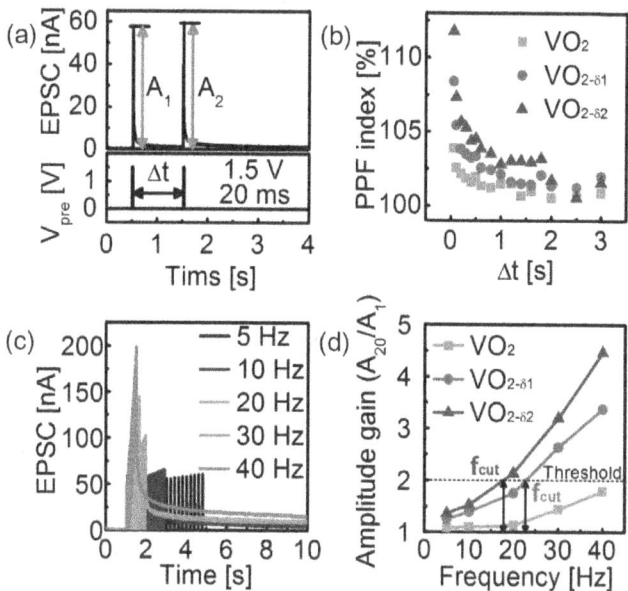

Figure 3.19 (a) EPSC event due to consecutive voltage stimulus separated in time; (b) PPF as a function of time interval between voltage pulses for different stoichiometries of VO_2; (c) PTP due to 20 consecutive pulses of amplitude 1.5 V and duration of 20 ms in an oxygen-deficient VO_2 transistor; and (d) EPSC amplitude gain as a function of pulse frequency for different devices [31]. Adapted with permission from [31]. © 2020 Wiley-VCH GmbH.

It is not essential to limit to just ionic liquids to observe this chemical interaction, even solid dielectrics that are ion conductors can serve as reservoirs for ions to pass through under an electric field. One example is shown in Figure 3.18 that illustrates the use of porous silica as a gate dielectric [31]. Vanadium dioxide films with varying density of oxygen vacancies were created by synthesis protocols. The oxygen vacancies tend to de-stabilize the insulating ground state resulting in increased conductance at room temperature and further enhance the rate of proton diffusion. The oxygen-deficient VO_{2-d} further serves as a synapse by modulating its resistance due to ion drift under successive electric fields. Figure 3.19 shows synaptic characteristics determined experimentally from such ion-gated transistors [31]. Figure 3.19(a) shows the excitatory postsynaptic current (EPSC) measured from the device for two consecutive voltage pulses and marked as A_1 and A_2. Paired-pulse facilitation is a property noted in biological synapses wherein the weight (or resistance in this case) is modified by a memory of the prior stimulus. In Figure 3.19(b), the PPF is shown as a function of time interval between voltage stimulus. It is clearly seen that as time interval increases, the PPF index (defined as PPF index = $A_2/A_1 \times 100\%$) is reminiscent of biological synapse behavior. Another aspect of interest is that the largest

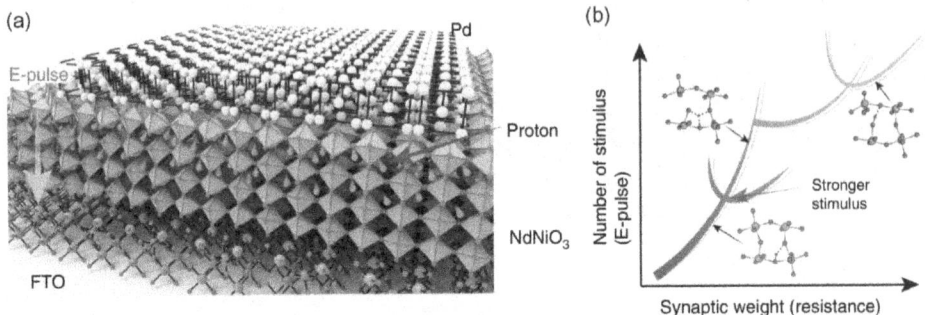

Figure 3.20 (a) Drawing of a Pd-NdNiO$_3$-FTO sandwich stack operable as a synapse. The hydrogen is doped into the NdNiO$_3$ film by gentle forming gas anneal. (b) Complex tree-like resistance states can be observed in such devices. The inset shows different configurations of protons in the lattice [32]. Adapted from [32], under the Creative Commons Attribution 4.0 license.

PPF was observed in the most oxygen-deficient sample. Figure 3.19(c) shows the peak EPSC after 20 consecutive pulses as a function of frequency and the amplitude gain is shown in Figure 3.19(d). The trend of increasing gain with increasing oxygen deficiency is seen. Such characteristics can be used to control the memory timescales to remember past stimuli and used in neural networks for learning and pattern recognition.

In the class of perovskite oxides, rare-earth perovskite nickelate systems such as SmNiO$_3$, NdNiO$_3$, and LaNiO$_3$ have emerged as an interesting candidate to design synapses. Upon hydrogen doping the nickelate, the resistance increases by several orders of magnitude at room temperature [32]. The hydrogen exists as a proton in interstitial sites while the electron donated from the hydrogen anchors to the Ni-O orbital manifold. By applying electric field stimulus, the protons can be shuffled inside the lattice. One hydrogen per unit cell doping results in about eight orders of magnitude change in resistivity indicating the extreme sensitivity to the dopant. Hence, small perturbations to the position and concentration of dopant can result in measurable changes in resistance. This can be elegantly exploited as a synapse in several device configurations. For instance, using a solid proton conductor such as (Ba,Y)ZrO$_3$, it is possible to design a three-terminal transistor that can modulate the channel resistance in a nonvolatile manner. Alternatively, using Pd or Pt catalytic electrodes, it is possible to locally dope the regions with a simple hydrogen anneal (since Pd and Pt are both excellent catalysts for hydrogen spillover), then use the device as a synapse. In this scenario, the device is a two-terminal tunable resistor. A schematic of H-doped NdNiO$_3$ film grown on FTO with a Pd top electrode is shown in Figure 3.20 [32].

Figure 3.21 shows an example of different resistance branches that can be experimentally obtained with voltage pulses. Figure 3.21(a) shows variation in resistance for fixed field strength while varying the pulse width. Figure 3.21(b)

3.4 Insulator–Metal Transition-Based Synapses

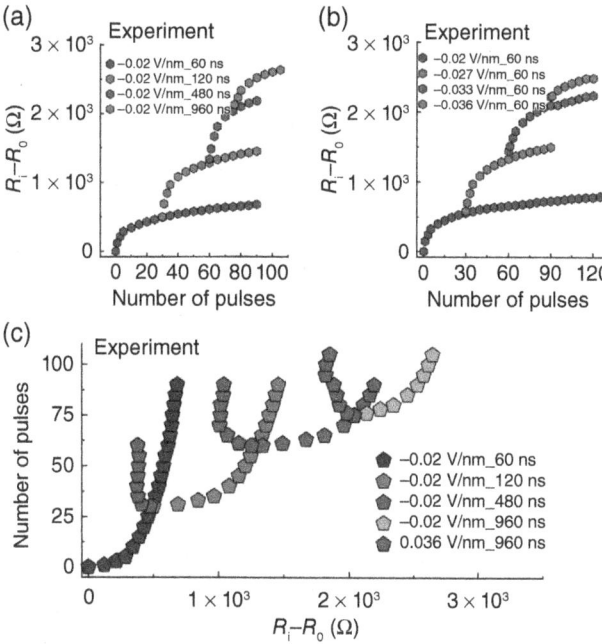

Figure 3.21 (a) Change in resistance as function of pulse number for fixed electric field strength and varying pulse width. (b) Change in resistance as function of pulse number for fixed pulse width and varying electric field strength. (c) Combination of pulse width and electric field values to demonstrate ultrametric tree-like memory states [32]. Adapted from [32], under the Creative Commons Attribution 4.0 license.

shows change in resistance for fixed pulse width of 60 ns while varying the field strength and (c) shows multiple branches of the resistance states while varying the polarity and strength of the electric stimulus [32]. By combining the sensitivity of the nickelate with the mobility of protons in the lattice, it is therefore possible to obtain large number of nonvolatile resistance states. It is also possible to obtain synaptic behavior and spike-timing-dependent plasticity in nickelates that contain oxygen vacancies [33]. Combining such devices with silicon electronics to demonstrate different types of learning mechanisms have been reported as well as demonstrating their potential in neuromorphic computing [34]. Using ionic liquids typically results in slow switching speeds in the range of several seconds to minutes or longer due to the mass transfer limitations across from the liquid to the solid phase and the charging timescales [35]. With catalytic electrodes in two-terminal solid-state devices, it is possible to tune resistance with nanosecond timescale pulses, resulting in high-speed, dense, all-solid state memory and low power operation.

An important point about the synapses described in this section is that the phase transition-driven conductance change is non-filamentary. Hence, a dopant front is being driven across the electrode–channel interface by electric fields. Although discussion here is limited to oxygen vacancies and protons, the reader should be aware that other small ions such as lithium ions could be inserted and removed by electric fields using suitable electrodes. In such cases, the ion radius and the energy barrier to migrate in a lattice determines the structural robustness during the intercalation process as well as electrical energy needed per synaptic weight update. To compete with biological synapses that consume about ~1 fJ per update, it is therefore essential to work with suitable ionic dopant species and host lattice for both easy migration and enough sensitivity in electronic structure to induce resistance modulation per electric pulse input. An area of future research is understanding the numerous metastable states that are formed at different doping concentrations. Additionally, randomness in impurity site occupancy in host lattice could open opportunities for design of stochastic synapses [36, 37].

3.5 Organic Materials-Based Synapses

Organic materials offer several unique features compared to their inorganic counterparts including lower cost of manufacturing via scalable deposition techniques, mechanical flexibility, and ability to create complex phase mixtures by wide choice of elemental incorporation into the polymer structural units. Another intriguing aspect of polymers is the potential to implant inside the body of an animal owing to their biocompatibility and direct interfacing with biological neural circuits for brain–machine interfaces. To date, a good fraction of artificial synapses demonstrated with organic materials rely on ion migration-driven conductance changes and/or structural rearrangements induced by Joule heating. Consequently, both two-terminal and three-terminal devices can be realized with these material systems. In a two-terminal device, one electrode serves as a presynaptic membrane, while the second emulates the postsynaptic membrane monitoring changes in conductance of the active channel. In a three-terminal device on the other hand, the gate electrode applies voltage inputs that emulate the presynaptic inputs and the drain electrode measures the conductance change of the channel across the source–drain channel region. A two-terminal device is perhaps simpler to fabricate and integrate in crossbar arrays used for weighted sum measurements. Three-terminal devices resemble a transistor structure and could offer additional degrees of freedom such as spatial summation of signals or mimic multisensory inputs if multiple gate electrodes are patterned. In this section, we discuss representative examples of synaptic functions based on conductance modulation. Please note that FE organic synapses are discussed in the chapter considering different classes of FE devices. A comprehensive discussion of organic systems for neuromorphic computing can be found elsewhere [38].

3.5 Organic Materials-Based Synapses

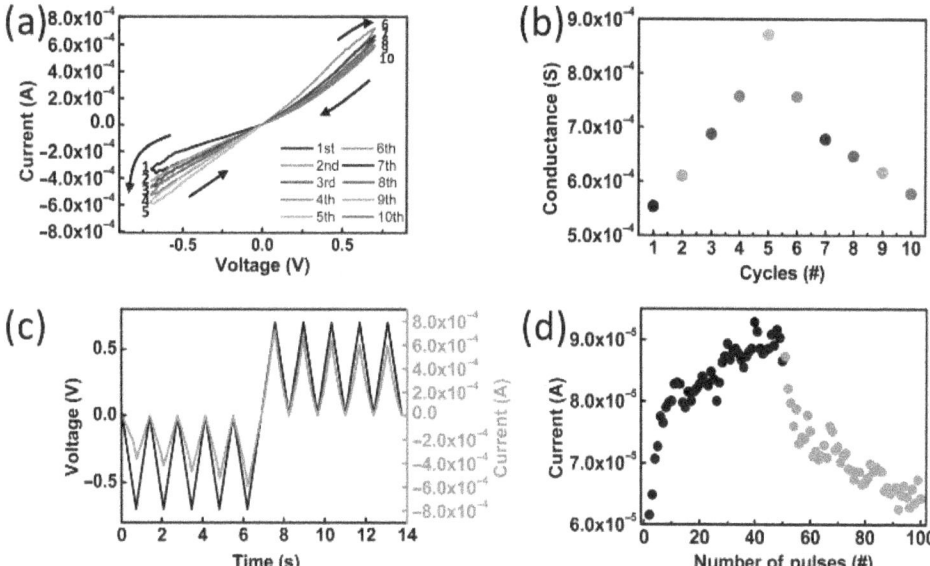

Figure 3.22 Electrical properties of a lignin-based device. The lignin film was deposited by spin coating and is approximately 100 nm thick. (a) I–V sweeps showing nonlinear response with increasing voltage. Five consecutive positive and negative cycles were measured. The magnitude of the current flowing through the device is dependent on sweep cycles. (b) Change in conductance during consecutive sweeps. The continuous increase and decrease in conductance is clearly seen. (c) Current response with time is shown in (c) indicating a gradual change in device conductance as a function of applied bias. Essentially this represents analog synaptic weight update. (d) Application of 50 consecutive programmed pulses (−0.7 V, 100 ms; +0.7 V, 100 ms) of negative and positive polarities, respectively, to emulate potentiation and depression characteristics [40]. Adapted with permission from [40]. Copyright © 2017 American Chemical Society.

We begin with a discussion of synapses fabricated with lignin, a polymer found in natural wood [39]. Lignin is a waste product of the paper industry and can host a variety of structures based on heat treatments thereby offering a simple knob to tune the electrical properties. The structure could be considered as a redox polymer comprising aryl propane units. Starting from a polyethylene terephthalate (PET) substrate, indium tin oxide (ITO) was deposited on top followed by lignin via spin coating. A top electrode comprising gold as an inert interface was then fabricated [40]. Figure 3.22 summarizes the basic electrical properties of the device.

The I–V sweeps show nonlinear behavior as well as history dependence. These suggest that the device region is evolving during the measurements. The conductance of the device can be increased/decreased incrementally with bias cycles. This is further confirmed in Figure 3.22(c) showing gradual changes in current flowing through the channel over multiple voltage bias pulses. Once the fundamental

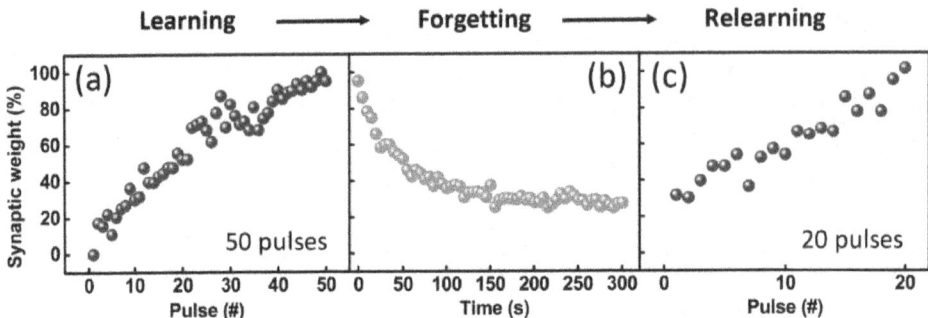

Figure 3.23 Synaptic weight increase as a function of pulse number for a total of 50 electric pulses, (b) spontaneous forgetting or decay of the resistance state, and (c) gradual increase of the weight at the onset of forgetting. In (c), it can be noted that only around 20 pulses were needed to bring the device to the original weight as opposed to 50 pulses for the first set of experiments [40]. Adapted with permission from [40]. Copyright © 2017 American Chemical Society.

electrical characteristics are obtained, with the right choice of device geometry and electrical pulse protocols, it is possible to examine host of synaptic functions including PPF, learning, forgetting mechanisms and other forms of plasticity. Figure 3.22(d) shows potentiation–depression curves for sequential application of voltage stimuli, which represents plasticity and is used as weight update inputs for neural networks. Another example is presented in Figure 3.23, demonstrating the learning, forgetting, and relearning process. Application of sequential pulses results in gradual increase of synaptic weight, while removal of the stimulus results in gradual decay of the resistance state and relearning occurs by subsequent application of electric stimulus.

An interesting aspect of the lignin study was that the relearning occurred with fewer number of pulses after the forgetting phase, suggesting an analogy to the learning process in biological brains where the brain can grasp previously learnt (and partially forgotten information) in a faster manner. Similar experiments using different top electrodes indicated the conductance modulation was likely due to modification of the carbon atom configuration inside the lignin upon electrical stimulus. The magnitude of the stimulus dictates the extent of Joule heating, which can result in the formation of graphitic structures, and changes the device conductance. This process is metastable and the extent of modification depends on the strength of the stimulus, hence a transition from temporary short-term plasticity (or memory) to long-term retention is possible.

Conducting filaments utilizing metal ions have also been demonstrated like for the inorganic systems discussed in Section 3.2. By systematic variation of the filament dimensions, it is possible to smoothly tune the resistance states by application of voltage pulses, and also realize nearly linear weight updates during

3.5 Organic Materials-Based Synapses

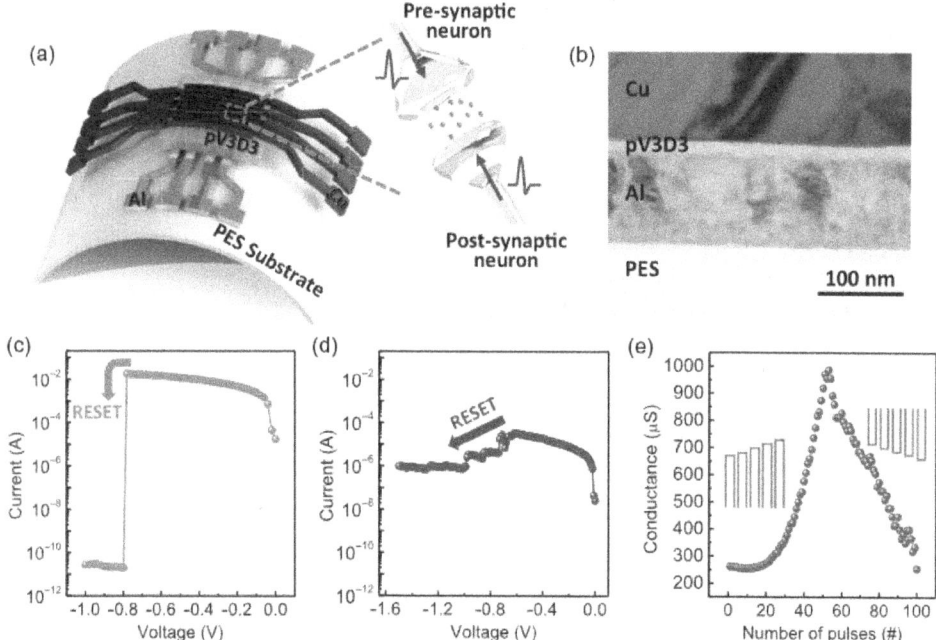

Figure 3.24 (a) Schematic of a pV3D3 polymer-based synaptic device fabricated with Cu and Al electrodes. The two electrodes represent the presynaptic and postsynaptic neuron terminals. (b) TEM image of the pV3D3 layer sandwiched between the two electrodes. (c) RESET behavior after filament formation and (d) RESET behavior as the compliance current was gradually changed. (e) Gradual change in conductance of the device as voltage pulse train was applied for both polarities (2–4 V, 60 ns pulses; −1.2 to −1.4 V, 100 ns, respectively) [41]. Adapted with permission from [41]. Copyright © 2019 American Chemical Society.

potentiation–depression measurements. Figure 3.24 shows a synaptic device fabricated with poly(1,3,5-trivinyl-1,3,5-trimethyl cyclotrisiloxane) (referred to as pV3D3). The polymer was deposited by chemical vapor deposition on Al film that served as a bottom electrode. The substrate was poly(ether sulfone) (PES) and the top electrode was copper [41]. The chemical stability of the polymer layer enabled use of photolithography to fabricate 5×5 μm^2 size lateral dimension cells. An electroforming process at the level of current compliance of ~10^{-5} A results in a SET process to form the copper conducting filaments. Re-sweeping the voltage without applying the current limit results in Joule heating that breaks the conducting pathway resulting in a RESET. By varying the compliance current levels gradually, it was possible to change the profile of the current–voltage relationships from gradual to abrupt switching, mainly through the dynamics of formation and rupture of filaments. Hence, it is quite clear how important managing the filamentary paths are in the case of organic devices too.

It is natural to wonder which electrode is primarily responsible for the formation of the conducting filaments if both are reactive. In this study, replacing the Al with Pt as bottom electrode still resulted in memory resistance switching behavior, thereby effectively ruling out the need for defective interfacial aluminum oxide or Al inclusions as the mechanism for the observed changes in resistance. Quantized changes in conductance can be further realized by delicate control over the size of the copper conducting filaments. By maintaining nearly atomically thin filaments, it is possible to realize quantum effects noted in nanoscale metals.

Besides the two-terminal devices described earlier, the insertion of ions or trapping/detrapping of charge carriers using a gate electrode stack in a three-terminal device can be an effective strategy for building synapses. For instance, using fullerene-PMMA composites as a tunneling layer in a transistor stack, it has been shown that the equivalent of a floating-body transistor can be designed with ON/OFF ratio approaching 1,000 along with synaptic functions including plasticity and transition from short-term to long-term plasticity. The fullerene domains embedded in the PMMA dielectric trap charge effectively and enable shift in threshold voltage of a transistor with pentacene as channel [42]. Hence, the three-terminal device acts as a synaptic transistor, where the gate electrode and channel synergistically provide learning and memory mechanisms. PEDOT:PSS polymer has been

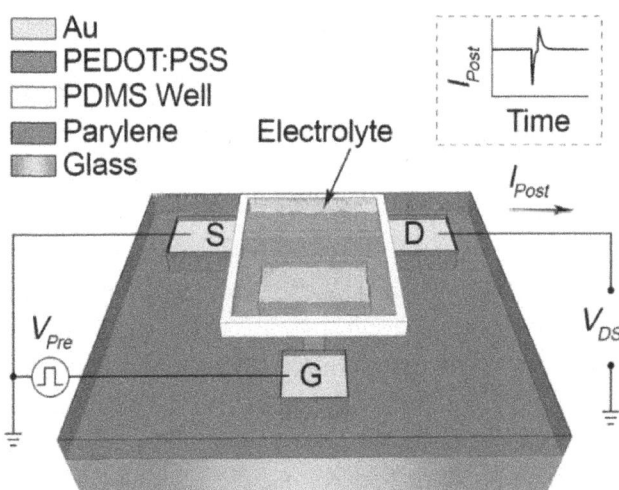

Figure 3.25 Schematic of an electrolyte-gated electrochemical transistor using PEDOT:SS as a semiconductor channel fabricated on glass substrate. The channel conductance is highly sensitive to potassium ion injection from the KCl electrolyte. Injection of K+ ions from the electrolyte results in de-doping of the channel increasing the resistance. The ions return back to the electrolyte upon removal of stimulus with a characteristic timescale [43]. Adapted with permission from [43]. © 2015 WILEY-VCH Verlag GmbH & Co. KGaA, Weinheim.

3.5 Organic Materials-Based Synapses

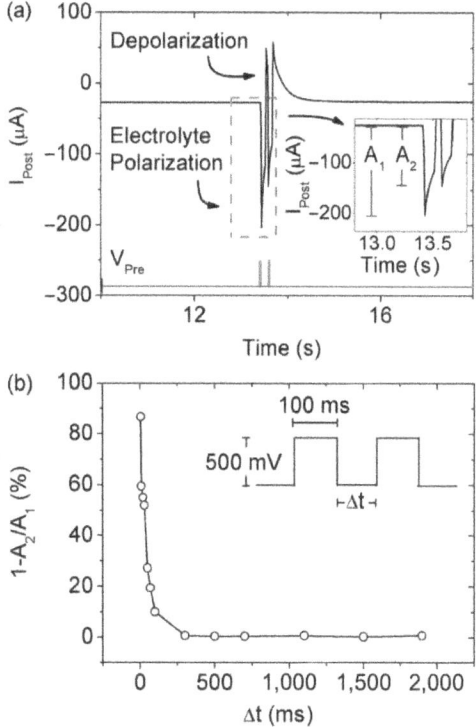

Figure 3.26 (a) A pair of presynaptic voltage pulses is applied to the gate electrode in contact with the KCl electrolyte. The postsynaptic drain current is measured as a function of time. The timescale is relevant due to the diffusional nature of the cations moving back and forth from the electrolyte into the polymer channel. The amplitude is A and is dependent on the voltage history. (b) Percentage of depression of A as a function of time interval between consecutive stimulus. Clearly, a significant decrease in the amplitude difference is seen as the time interval between subsequent pulses is increased as the channel relaxes closer and closer to equilibrium state [43]. Adapted with permission from [43]. © 2015 WILEY-VCH Verlag GmbH & Co. KGaA, Weinheim.

explored as a synaptic material by interfacing with KCl electrolyte [43]. Upon application of a voltage bias, the potassium ions enter the polymer resulting in de-doping and consequently a change in resistance (Figure 3.25).

The ions return back to the electrolyte upon removal of the stimulus. This offers an elegant means to create different forms of plasticity: by understanding the relevant timescales for ion injection and removal from the polymer layer, it is possible to design voltage pulse profiles to demonstrate different learning mechanisms. Figure 3.26 shows an example of pair-pulsed depression in this device. Voltage pulse pairs of magnitude 500 mV was applied consecutively with time interval varied from 1 to 1,900 milliseconds for a constant source-drain bias of −300 mV. From Figure 3.26(a), A2 is less than A1, that is the postsynaptic current A2 for the second voltage is depressed compared to A1. Interestingly, this difference is a function of the time interval between the pair of pulses being applied. This is shown in Figure 3.26(b) where a sharp decrease in the amplitudes is seen as time interval is increased to 1,900 ms. In fact, above a certain timescale around 200 ms, the depression is nearly absent. This high frequency paired pulse depression is reminiscent of the short-term depression that was discussed in Chapter 1. While in biological neural circuits, the depression is attributed to the decrease of ability of neurotransmitter containing vesicles to release signaling molecules, in the polymer transistor it pertains to the inability of cations to return to the electrolyte. This results in polarization build up at the polymer channel-gate interface and short-term depression. A deep understanding of ion dynamics at the polymer interface with ionic electrolyte gates allows emulation of important synaptic behavior found in biological neural circuits. We point the reader to related articles for further reading about different polymers that can be utilized for such applications [44–46].

3.6 Two-Dimensional and Layered Material Synapses

There has been an explosion in research in the areas of 2D and layered semiconductors in past couple of decades since graphene burst into the scene. Since then, beyond single elemental materials (e.g., C, P, and Te) being explored at monolayer or few layer geometries, a plethora of compounds (e.g., MoS_2, TaS_2, MoO_3, h-BN, Ti_3C_2, and $MoTe_2$) have been realized in freestanding forms. Often, they are studied as flakes that are anywhere from few to several atomic layers thick in the vertical direction and can be placed on arbitrary substrates. The thickness of these systems can be controlled systematically from single atom thick (monolayer) to several atomic layers. One characteristic that appears common across these systems is the property that intralayer bonding energies are much stronger than out-of-plane interactions, hence they are also referred to as van der Waals systems. This characteristic enables quick fabrication of heterostructures by exfoliation and stacking dissimilar materials for proof-of-concept studies. In recent years as the studies show promising properties for practical technologies, methods for scalable synthesis and monolithic integration with traditional electronic platforms such as silicon wafers are also being investigated. In this section, we discuss representative examples of synaptic devices realized with such 2D materials as the active layer.

To realize synaptic properties, it is necessary to have both short-term and long-term memory as well as plasticity. Hence, the basic principles to demonstrate artificial synapses with these systems are somewhat similar to bulk (i.e., 3D) semiconductors. Control of ion diffusion by electric stimuli, modification of structural symmetry via strain or electric field gating that results in

3.6 Two-Dimensional and Layered Material Synapses

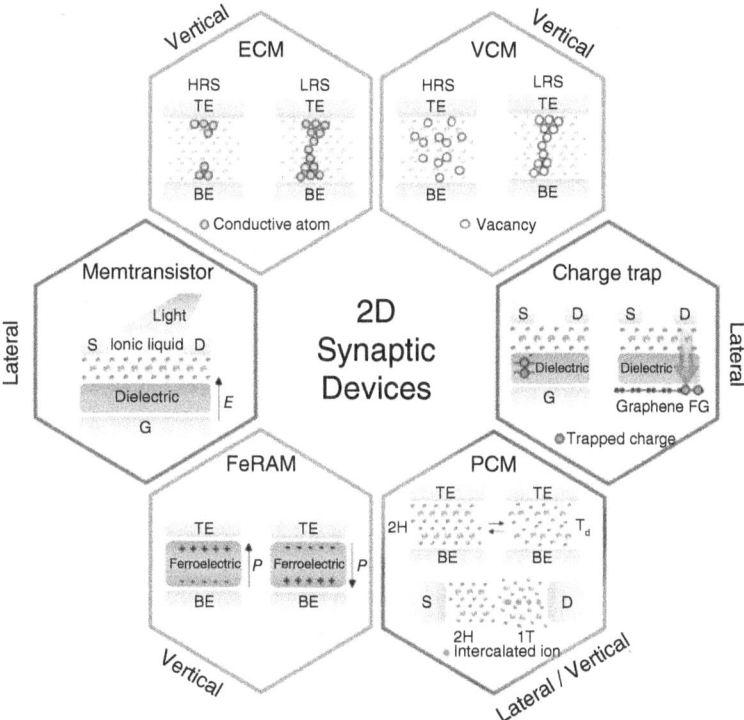

Figure 3.27 A vast array of 2D semiconductors have been explored in literature for artificial synapses. The mechanisms include cation migration under electric fields (ECM), vacancy migration (VCM), trapping and detrapping of charges similar to flash memory, three-terminal memory transistors (referred to as memtransistors), phase-change (PCM) layered systems, and ferroelectric systems (FeRAM). The device geometry could be lateral or vertical, that is relying on in-plane or out-of-plane conduction, respectively. In the figure, TE and BE refer to top and bottom electrodes, S and D refer to source and drain, respectively [47]. Adapted with permission from [47]. © 2021 Wiley-VCH GmbH.

different band gaps for electrical conduction or light propagation, introduction of trapping sites in the semiconductor stack, and tunnel junctions are all viable approaches explored in literature. Figure 3.27 summarizes various mechanisms that have been utilized to demonstrate artificial synapses with 2D systems [47]. We discuss below how the reduced dimensionality of the semiconductor channel could offer some unique advantages with specific examples. Besides the obvious challenge of developing wafer-scale fabrication technologies for such emerging semiconductors, we consider potential limitations when the layer thickness approaches one or few atomic units.

We begin with representative discussion on electrochemical metallization-based synapses using boron nitride sheets. Conducting filaments formed across a

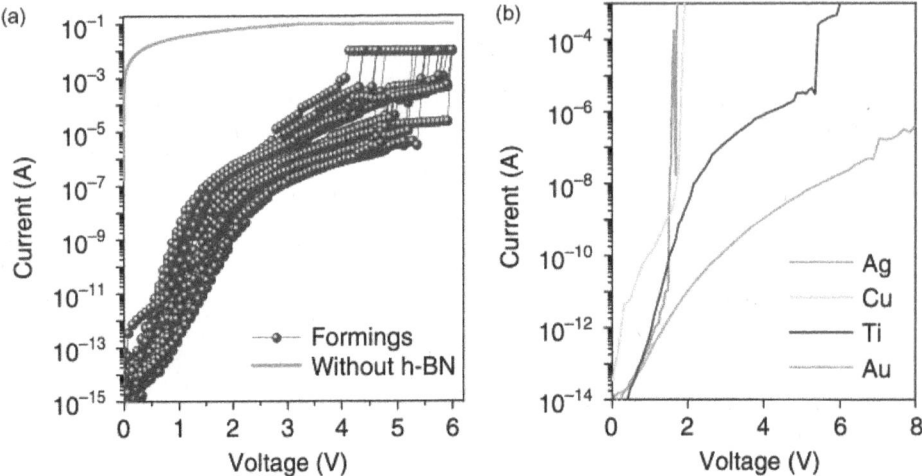

Figure 3.28 (a). Current–voltage characteristics taken from Au-Ti/h-BN/Au sandwich stack indicating gradual rise in current followed by breakdown under high electrical stress. (b) Electrode dependence of current rise in junctions. The nature of defects formed in the BN and metal conducting filament formation that is needed for the breakdown clearly are both dependent on the choice of the metal electrodes [48]. Adapted with permission from [48]. Copyright © 2018, The Author(s).

semiconductor channel due to oxidation–reduction reactions at the electrode interfaces offer a path to designing synapses. Active electrodes include copper, aluminum, or silver that can undergo redox reactions forming ions that respond to electric fields, while inert metals such as gold serve as counter electrodes. As one example, boron nitride sandwiched between metal electrodes can function as tunable memory devices as shown in Figure 3.28(a) [48].

Few-layer thick hexagonal boron nitride sheets were grown by chemical vapor deposition onto copper foils followed by top electrode deposition. Figure 3.28(a) shows forming characteristics in 5 μm × 5 μm device junctions (15 different ones) with Au-Ti and Au electrodes separated by 15–18 layers of hexagonal boron nitride. It is clearly seen that a gradual rise in current followed by abrupt increase signals the breakdown event capped at 10 mA current limit. Interestingly, in samples without the BN layers, there is no such behavior (red curve) indicating the role of the 2D layers. Further, as seen in Figure 3.28(b), the choice of the electrode significantly influences the electrical characteristics owing to their redox characteristics. The mechanism of electrical switching stems from native defects in the insulating layer that are formed during the deposition. Metal penetration into the nitride layer along with vacancies created due to B diffusion toward the anode appears to result in the sharp switching of the conducting states. The filament mechanism enables demonstration of basic synaptic functions such as PPF, spontaneous relaxation akin to Ebbinghaus forgetting, and bias-dependent

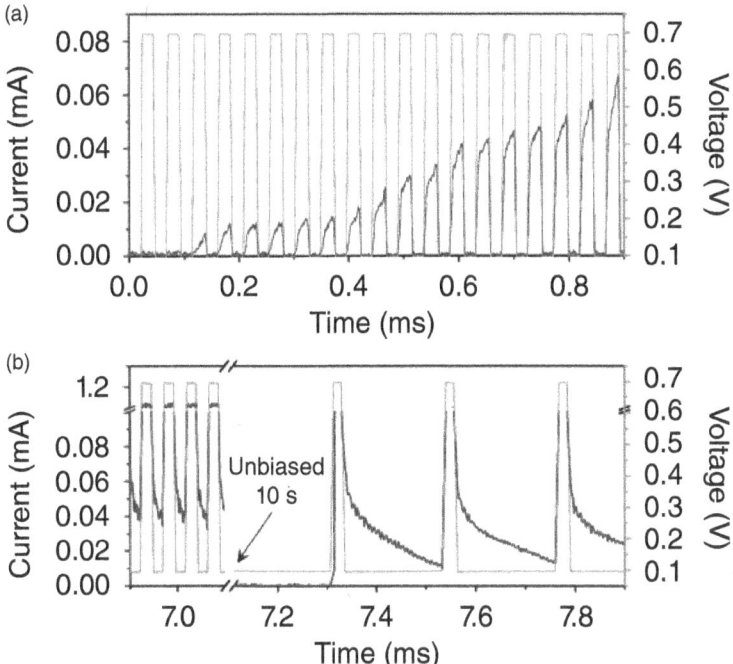

Figure 3.29 Synaptic action with h-BN channels. (a) Voltage dependence of current across h-BN junction. (b) Relaxation of current across the junction when the voltage pulse is off for 10 s. The current decays back toward initial state due to diffusion of cations. The electric field strength controls the cation concentration inside the BN layers. Further, the greater the dopant distribution inside the BN layer, the longer it takes to decay to original state spontaneously [48]. Adapted with permission from [48]. Copyright © 2018, The Author(s).

memory retention timescale. Figure 3.29(a) and (b) shows examples of data measured from 5- to 7-layer thick h-BN sheets [48].

We note here that while the operation of h-BN junctions is similar to other conducting bridge filamentary switches, there also exist subtle differences when contrasted against metal oxides. This is because the remaining nitrogen is not conductive enough (unlike reduced cations in an oxide dielectric) to support electron transport. Therefore, diffusion in and out of metal ions from the electrodes is an essential part of the memory switching and retention process. Hence, by using an anion framework that limits the vacancy-mediated conduction, it is possible to exploit the redox capability of the electrode material to control the overall synapse characteristics and related voltage–current parameters. Point defects in 2D systems are beneficial to realizing synaptic function and has been studied in MoS_2 in some detail. Sulfur vacancies in this system can enable conduction by forming filaments. Hence, formation and rupture of the filaments are natural mechanisms to realize nonvolatile memory. It is however challenging to control the arrangement of these filaments due to their miniscule size. Manipulation of local defects are often performed with a scanning tunneling microscope and restricted to mechanistic studies. Variability from C2C and D2D therefore represent major challenges in this class of devices. In almost all filamentary switches, the unpredictable nature of filament formation renders the technology subject to large variability. While still in early stages, there are efforts to mitigate this formidable challenge. Strategies include interface and defect engineering such as use of a silicon interfacial layer on MoS_2 that can reduce the C2C variability by ten times [49]. The use of polycrystalline layers is another approach explored in literature. The presence of grain boundaries in polycrystals results in extended planar defects that obviate the need for forming process simplifying device operation.

Ferroelectric devices were discussed in Section 3.3 wherein controlling the relative fraction of domains with certain polarization orientation resulted in analog memory. We will also discuss in Chapter 4 on artificial neurons how lossy ferroelectrics can lose the polarization with time due to leakage currents. In the case of devices that incorporate 2D semiconductors, there are a few different schemes to utilize the property of ferroelectricity. In one approach, an FE can be directly integrated onto a 2D channel as gate. The enhanced charge accumulation characteristics of the FE layer can be utilized to modify the nature of charge carriers in the 2D channel thereby obtaining both memory- and field-dependent conductance. This has been demonstrated in graphene channels with polyvinylidene fluoride (PVDF) FE stack [50]. Alternatively, one can explore emerging 2D systems that show FE properties, examples include $CuInP_2S_6$ (CIPS). When considering such ultra-thin ferroelectrics, it is suited to use them in tunnel junctions wherein the orientation of polarization controls the tunnel current across the junction. CIPS-based junctions comprising Ti and Au electrodes have demonstrated synapse functions such as potentiation, depression, PPF, plasticity, and encoding information in simple 3 × 3 device arrays [51]. Given the rapid pace of discovery of novel 2D materials across various materials systems, it will be interesting to see how this field evolves in the coming years. Properties of specific interest to electronic synapses include polarization retention over extended periods of time to realize long-term plasticity, control over domain population distribution to realize multiple resistance states, and switching endurance in junctions with CMOS-compatible electrodes.

It is worth noting that phase changing 2D systems are another class of interesting candidates to demonstrate synapses. Structural transformations between phases that possess different electronic transport gaps is one of the most studied routes. As an example, MoS_2 can be semiconducting or metallic depending on the crystal structure. The 2H phase is semiconducting (with trigonal prismatic structure) while the distorted octahedral 1T phase is metallic. By electrochemical insertion of lithium into the MoS_2 lattice, it is possible to drive the local phase change by electric field control of the dopant distribution between electrodes [52]. Such redistribution of the lithium ions not only enables synaptic function but also emulates synaptic competition and cooperation effects in neighboring interconnected devices (due to enrichment or depletion of lithium dopants) that has been noted in biological neural circuits. Hence, controlling spatial distribution of injected dopants mimics neurotransmitter diffusion in biological networks that can affect synaptic weights in a small region of the network. Ionic gating is another effective method to realize synaptic function in 2D systems. For instance, in the same material system of MoS_2, using lithium perchlorate as an electrolyte, modulation of conductance of the semiconductor to realize potentiation and depression has been achieved [53]. The high capacitance of the electrolytic gate is well-suited to modify carrier density in the ultra-thin channel as the field penetrates an appreciable thickness fraction of the channel. Furthermore, beyond a certain threshold field, the ions can also penetrate the channel enabling nonvolatile conductance change. It is reasonable to conclude that the low dimensionality of the channel offers flexibility in terms of electric field screening as well as short diffusion distances to induce measurable changes in the conductance. It is important to note however that the dynamics inherent in such devices are quite limited, offering response in the milliseconds timescale and slower. Further increase in response speed is either not possible due to the sluggish nature of lithium-ion motion, electrolytic capacitor charging time constant or potential damage to the crystal lattice by ion trapping. We note here that the structural phase transitions can also be accomplished by thermal effects like the well-known 3D counterparts such as GST. Voltage pulses that result in resistive heating of the local regions is another route to realize two-terminal synapses with layered semiconductors [54]. The low thermal mass can result in rapid heating and hence fast phase changes, while the return to the original state will be limited by thermal dissipation and undercooling (hysteresis) effects. The thermal time constants and device geometry will determine the timescales for short-term plasticity. We point the reader to recent reviews that delve into synapses realized with various 2D and Van der Waals semiconductors for further studies [55, 56].

3.7 Spintronic Synapses

The basic device structure underlying ferromagnetic synapses is the magnetic tunnel junction (MTJ) which consists of two nanomagnetic layers (like CoFe or CoFeB) separated by a tunneling oxide like MgO [57, 58]. A nanomagnetic layer can be stabilized along two collinear but oppositely directed magnetization states in the absence of a stimulus while an external magnetic field or current can be used to manipulate the state of the magnetic layer. As shown in

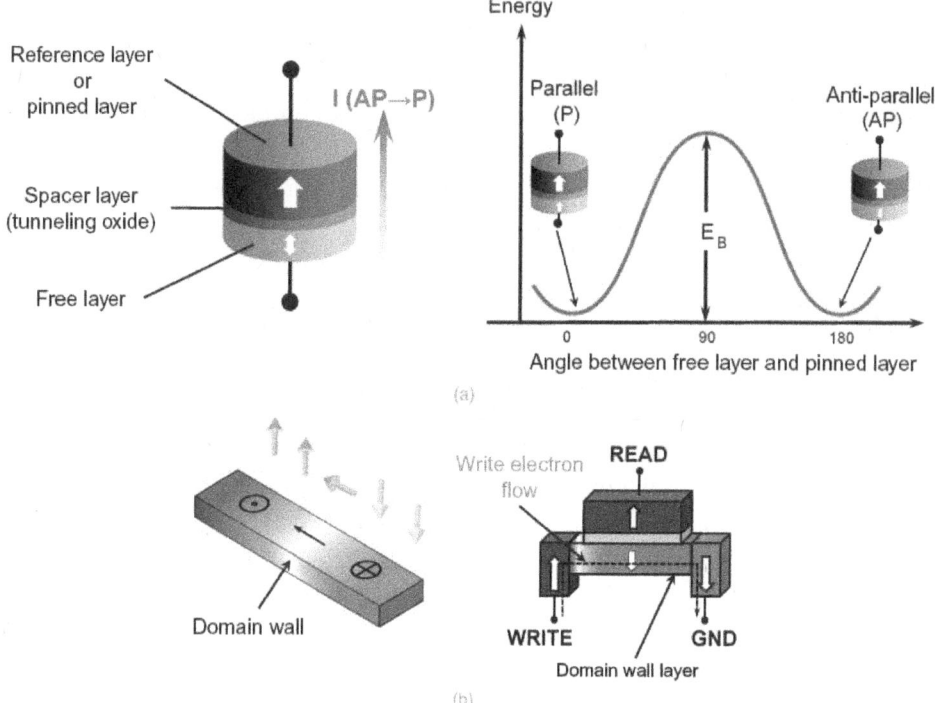

Figure 3.30 (a) MTJ device structure. The device can be switched from AP to P state by passing current from the free to the pinned layer and vice versa. The two extreme states are stabilized by an energy barrier that is dependent on the anisotropy and volume of the magnet. (b) Multiple resistive states can be achieved in an MTJ structure where the free layer has an elongated shape to stabilize multiple domains in the device. While the device state is read through the tunneling junction, the domain wall location is programmed by passing current horizontally through the free layer [57]. Adapted with permission from [57]. IOP Publishing © 2018 The Japan Society of Applied Physics.

the MTJ structure in Figure 3.30(a), one of the magnetic layers is "hardened" or "pinned" to have a fixed magnetic orientation (referred to as pinned layer) while the other layer is switched to different orientations (referred to as free layer). Depending on the relative magnetization orientations of the two layers, the device can exhibit a variable resistance. The two extreme resistance states of the device are usually referred to as the parallel (P: when both layers' magnetizations are in the same direction) and anti-parallel (AP: when both layers' magnetizations are in the opposite direction) orientation where the AP resistance state is higher than the P state.

While magnetic memories with bistable states are commercially available as STT-MRAMs (STT stands for spin-transfer torque wherein the magnetic layer is switched by torque exerted by an external current; MRAM stands for magnetic

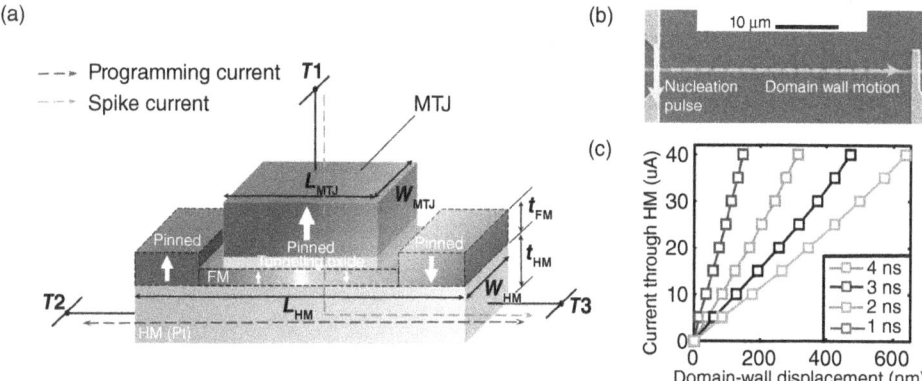

Figure 3.31 (a) Spintronic synapse where the programming current flows through a ferromagnet-heavy metal bilayer structure [61]. (b) Scanning electron micrograph of domain wall motion in a nanowire where the left current pulse is used to nucleate a domain wall. It is then driven to the right by an applied current [62]. (c) Variation of domain wall displacement with programming current amplitude for different pulse durations in a CoFe (0.6 nm)-Pt(3 nm) bilayer structure with 20 nm cross-sectional area [61]. Adapted with permission from [61] and [62]. (a and c) © 2016 American Physical Society; (b) Copyright © 2013, Springer Nature Limited.

random access memory) today, multiple nonvolatile programmable states are required to realize a synaptic device. Such a functionality can be achieved in a device structure as shown in Figure 3.30(b) where the free layer is elongated to stabilize multiple magnetic domains in the device with a domain wall separating the different magnetization regions. Depending on the relative proportion of magnetic domains in the free layer, the device exhibits a variable conductance. The conductance can be programmed by the amplitude and duration of the programming current flowing through the device. The domain wall movement direction also depends on the direction of the programming current, thereby enabling the realization of both potentiation and depression operations.

Various emerging physical phenomena have been leveraged in these spintronic devices for efficient manipulation of the device state such as voltage controlled magnetic anisotropy [59] and spin Hall effect [60], among others to reduce operating voltage and current levels. Let us consider a particular device example in Figure 3.31 where a magnetic synapse is represented by a domain wall motion-based device with a bilayer stack consisting of a ferromagnet lying on top of a heavy metal [61, 62]. In such stacks, an input charge current flowing through the heavy metal causes repeated scattering of incoming in-plane spin polarized electrons at the interface of the two layers, thereby transferring multiple units of spin angular momentum to the ferromagnet lying on top. Charge current flowing between terminals T2 and T3 through the heavy metal layer can program the

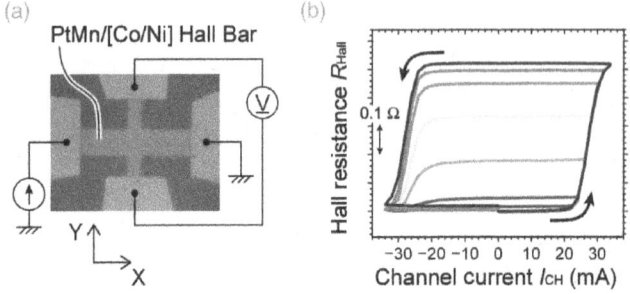

Figure 3.32 (a) An AFM/FM heterostructure-based Hall bar device [64]. (b) Measured Hall resistance with varying channel current. The different colored lines correspond to different maximum channel currents during the hysteresis sweep [64]. Adapted with permission from [64]. IOP Publishing © 2017 The Japan Society of Applied Physics.

domain wall position and thereby modulate the conductance measured between terminals T1 and T3. Such a decoupled read–write three terminal device structure has multiple advantages at the system level. These will be discussed in Chapter 6 in further detail. However, few key points worth remembering regarding the device operation are that these are magneto-metallic devices requiring very low currents for switching where the input write current flows through a low resistance heavy metal layer (instead of the tunneling junction barrier). The critical current required for domain wall displacement depends on the width of the device while the length dictates the number of available programmable states due to more pinning locations available for the domain wall. Additionally, to stabilize domain wall pinning in the presence of defects, notches can also be used along the nanotrack.

Innovations in the material stack have also been pursued to implement multilevel programming capability in alternative spintronic device structures. Recent studies on antiferromagnet/ferromagnet (AFM/FM) bilayer systems (for instance, PtMn/Co-Ni) have demonstrated sufficiently large SOT to switch an FM layer lying on top in the absence of any external field due to the exchange bias of the AFM layer [63]. In addition, the amount of reversed magnetization in such AFM/FM systems can be tuned in accordance to the magnitude of the applied write current, thereby leading to multilevel nonvolatile device states – suitable for neuromorphic computing applications. Figure 3.32 shows a device structure with a ferromagnetic layer lying on top of an AFM channel processed into Hall bar structures [64]. Input write current, I_{CH}, flowing through the device will result in a change of the anomalous Hall resistance, R_{Hall}, thereby emulating synaptic functionality.

As devices are scaled down, mono-domain magnets lose their nonvolatile properties, and their state may start decaying in the absence of a current stimulus. This is due to damping and precession torques that are present in a magnet whose

3.7 Spintronic Synapses 69

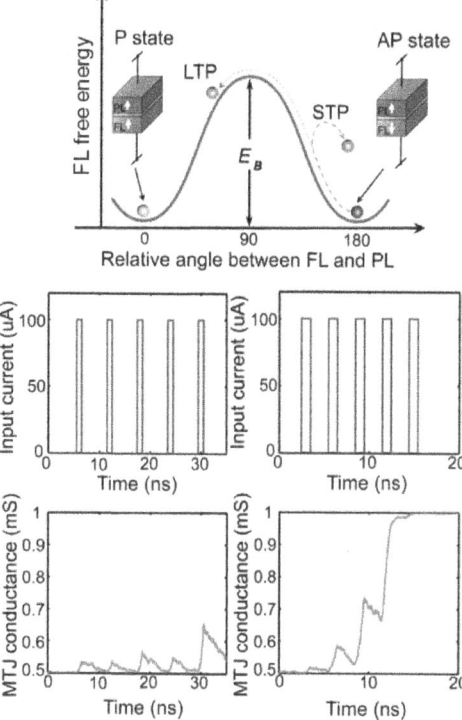

Figure 3.33 Frequency-dependent synaptic learning where a stimulus with higher frequency causes the device to switch to the other stable state [67]. Adapted with permission from [67]. 2017 © AIP Publishing.

magnetization is not present along the stable easy axis. The device will therefore try to stabilize along the easy axis direction when no current is applied (see Figure 3.33). Such a property can be utilized to design short-term plasticity and long-term potentiation based learning in spin synapses where a more frequent current stimulus will cause the device to transition to the other stable state (long-term potentiation) while a less frequent one will cause plasticity changes in a smaller timescale (short-term plasticity) [65].

Spin textures other than domain walls have also been used in multiple works. One example of a novel spin texture is a skyrmion which is a topologically protected two-dimensional texture that can be of small size (10–20 nm laterally). The motion and displacement of skyrmions can be controlled by electric currents. Therefore, similar to domain walls, skyrmions can be treated as the state variable and used to construct synapses and neurons [66]. However, modulating the number of injected skyrmions and weak magnetoresistive readout has still limited its widespread applicability. In general, low ON–OFF ratio is still a concern for the broad category of MTJ devices and material and device innovations are being pursued to overcome this challenge [67]. Prototype experimental demonstration of a system of spin neuromorphic devices has already been pursued [68–70] but remain less mature than some of the other technologies such as resistive RAMs and phase change memories.

3.8 Photonic Synapses

While Sections 3.2–3.7 have discussed several device technologies, programming speed remains a major concern, especially for designing scalable neuromorphic architectures. Integrated photonics provides an alternative pathway to implement ultra-fast neuromorphic substrates. While different methodologies have been explored in this field [71–73], we will focus on one particular instance [74] of a nonvolatile implementation enabled by experiments showcasing sub-ns write operation in phase-change memory ($Ge_2Sb_2Te_5$ – GST) devices through optical pulses [75]. Figure 3.34 depicts a photonic synapse where a single-bus microring resonator has a GST device element embedded on top of it. The cross-sectional view is also shown in the figure to illustrate that the device is fabricated on top of SiO_2 substrate and consists of a rectangular and ring waveguide where the GST element is present on the arm of the ring waveguide [76].

Waves travelling in the rectangular waveguide are partially coupled to the ring and undergo constructive interference when the round-trip phase shift becomes a multiple of 2π. Under this resonant condition, the transmission T_λ is a function of the attenuation factor in the ring waveguide and self-coupling coefficient. Interestingly, this attenuation factor varies with the effective imaginary refractive

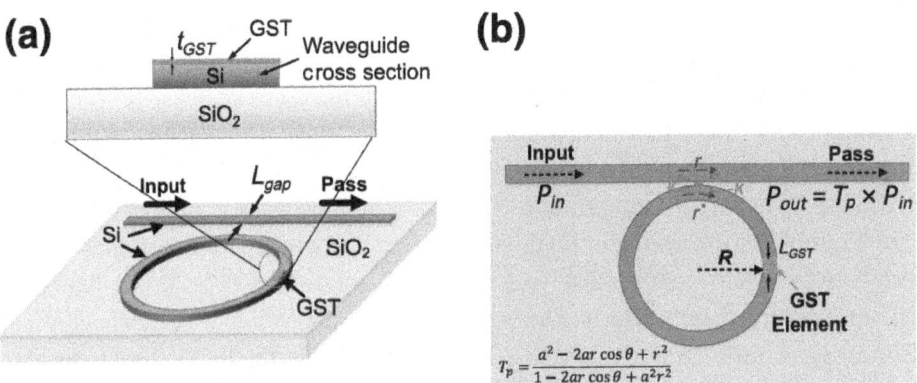

Figure 3.34 (a) A microring resonator with embedded GST serving as a photonic synapse. (b) Top view of the device structure where synaptic scaling operation is represented by the transmission at the resonant wavelength [77]. Adapted with permission from [77]. © 2019 American Physical Society.

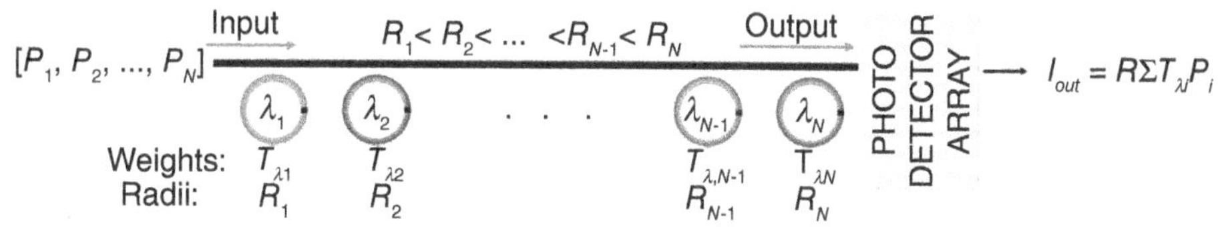

Figure 3.35 Photonic synaptic dot-product engine based on an array of microring resonators with gradually increasing radii representing various wavelengths [77]. Adapted with permission from [77]. © 2019 American Physical Society.

index of the GST material which is greatly different for GST in the amorphous and crystalline states. Therefore, by programming the GST in different partially crystallized states, one can vary the level of transmission at the resonant wavelength, T_λ. For an input optical pulse with power P_{in}, multiplicative synaptic scaling is reflected in the output power, $P_{out} = T_\lambda \cdot P_{in}$.

Wavelength division multiplexing can be used to implement the dot-product computing kernel. Figure 3.35 shows the photonic synapse dot-product engine [77] where multiple ring resonators of gradually increasing radius are used to represent multiple wavelengths. As we have seen for other synaptic technologies, array sizes for these dot-product engines are limited. For nanoelectronic devices switched by electric pulses, it was attributed to sneak path issues and parasitics. For such photonic synapses, the free-spectral range of the ring resonator limits the array dimensions. Further, each channel needs to have sufficient isolation to minimize interference effects. As shown in the figure, an input spike is provided through the input port of the array. The amplitude of each wavelength gets modulated by the corresponding GST element on the ring resonators. The output is therefore a multi-wavelength spike consisting of different $T_\lambda \cdot P_{in}$ terms for each wavelength. Photonic synapses therefore offer a competing paradigm for enabling high-speed neuromorphic computing systems.

References

[1] Xia, Q. and Yang, J. J., 2019. Memristive crossbar arrays for brain-inspired computing. *Nature Materials*, 18(4), pp. 309–323.

[2] Li, Z., Liu, F., Yang, W., Peng, S. and Zhou, J., 2021. A survey of convolutional neural networks: Analysis, applications, and prospects. *IEEE Transactions on Neural Networks and Learning Systems*, 33(12), pp. 6999–7019.

[3] LeCun, Y., Bengio, Y. and Hinton, G., 2015. Deep learning. *Nature*, 521(7553), pp. 436–444.

[4] Sengupta, A., Banerjee, A. and Roy, K., 2016. Hybrid spintronic-CMOS spiking neural network with on-chip learning: Devices, circuits, and systems. *Physical Review Applied*, 6(6), p. 064003.

[5] Rajendran, B., Liu, Y., Seo, J. S., Gopalakrishnan, K., Chang, L., Friedman, D. J. and Ritter, M. B., 2012. Specifications of nanoscale devices and circuits for neuromorphic computational systems. *IEEE Transactions on Electron Devices*, 60(1), pp. 246–253.

[6] Sengupta, A., Ankit, A. and Roy, K., 2017, May. Performance analysis and benchmarking of all-spin spiking neural networks (special session paper). In *2017 International Joint Conference on Neural Networks (IJCNN)* (pp. 4557–4563). IEEE.

[7] Chua, L., 1971. Memristor: The missing circuit element. *IEEE Transactions on Circuit Theory*, *18*(5), pp. 507–519.

[8] Nandakumar, S. R., Le Gallo, M., Piveteau, C., Joshi, V., Mariani, G., Boybat, I., Karunaratne, G., Khaddam-Aljameh, R., Egger, U., Petropoulos, A. and Antonakopoulos, T., 2020. Mixed-precision deep learning based on computational memory. *Frontiers in Neuroscience*, *14*, p. 406.

[9] Sun, X., Liu, R., Peng, X. and Yu, S., 2018, October. Computing-in-memory with SRAM and RRAM for binary neural networks. In *2018 14th IEEE International Conference on Solid-State and Integrated Circuit Technology (ICSICT)* (pp. 1–4). IEEE.

[10] Jerry, M., Chen, P. Y., Zhang, J., Sharma, P., Ni, K., Yu, S. and Datta, S., 2017, December. Ferroelectric FET analog synapse for acceleration of deep neural network training. In *2017 IEEE International Electron Devices Meeting (IEDM)* (pp. 6–2). IEEE.

[11] Saha, A., Islam, A. N. M., Zhao, Z., Deng, S., Ni, K. and Sengupta, A., 2021. Intrinsic synaptic plasticity of ferroelectric field effect transistors for online learning. *Applied Physics Letters*, *119*(13), pp. 133701-1–133701-6.

[12] Agrawal, A., Chakraborty, I., Roy, D., Saxena, U., Sharmin, S., Koo, M., Shim, Y., Srinivasan, G., Liyanagedera, C., Sengupta, A. and Roy, K., 2020. Revisiting stochastic computing in the era of nanoscale nonvolatile technologies. *IEEE Transactions on Very Large Scale Integration (VLSI) Systems*, *28*(12), pp. 2481–2494.

[13] Schmitt, S., Klähn, J., Bellec, G., Grübl, A., Guettler, M., Hartel, A., Hartmann, S., Husmann, D., Husmann, K., Jeltsch, S. and Karasenko, V., 2017, May. Neuromorphic hardware in the loop: Training a deep spiking network on the BrainScaleS wafer-scale system. In *2017 International Joint Conference on Neural Networks (IJCNN)* (pp. 2227–2234). IEEE.

[14] Sengupta, A. and Roy, K., 2016. Short-term plasticity and long-term potentiation in magnetic tunnel junctions: Towards volatile synapses. *Physical Review Applied*, *5*(2), p. 024012.

[15] Jeong, D. S., Kim, I., Ziegler, M. and Kohlstedt, H., 2013. Towards artificial neurons and synapses: A materials point of view. *RSC Advances*, *3*(10), pp. 3169–3183.

[16] Ohno, T., Hasegawa, T., Tsuruoka, T., Terabe, K., Gimzewski, J. K. and Aono, M., 2011. Short-term plasticity and long-term potentiation mimicked in single inorganic synapses. *Nature Materials*, *10*(8), pp. 591–595.

[17] Du, C., Ma, W., Chang, T., Sheridan, P. and Lu, W. D., 2015. Biorealistic implementation of synaptic functions with oxide memristors through internal ionic dynamics. *Advanced Functional Materials*, *25*(27), pp. 4290–4299.

[18] Kim, S. G., Han, J. S., Kim, H., Kim, S. Y. and Jang, H. W., 2018. Recent advances in memristive materials for artificial synapses. *Advanced Materials Technologies*, *3*(12), p. 1800457.

[19] Ielmini, D. and Waser, R. eds., 2015. *Resistive switching: From fundamentals of nanoionic redox processes to memristive device applications*. Weinheim: John Wiley & Sons.

[20] Wang, Z., Joshi, S., Savel'ev, S. E., Jiang, H., Midya, R., Lin, P., Hu, M., Ge, N., Strachan, J. P., Li, Z. and Wu, Q., 2017. Memristors with diffusive dynamics as synaptic emulators for neuromorphic computing. *Nature Materials*, *16*(1), pp. 101–108.

[21] Chanthbouala, A., Garcia, V., Cherifi, R. O., Bouzehouane, K., Fusil, S., Moya, X., Xavier, S., Yamada, H., Deranlot, C., Mathur, N. D. and Bibes, M., 2012. A ferroelectric memristor. *Nature Materials*, *11*(10), pp. 860–864.

References

[22] Böscke, T. S., Müller, J., Bräuhaus, D., Schröder, U. and Böttger, U. J. A. P. L., 2011. Ferroelectricity in hafnium oxide thin films. *Applied Physics Letters*, 99(10), pp. 102903-1–102903-3.

[23] Max, B., Hoffmann, M., Mulaosmanovic, H., Slesazeck, S. and Mikolajick, T., 2020. Hafnia-based double-layer ferroelectric tunnel junctions as artificial synapses for neuromorphic computing. *ACS Applied Electronic Materials*, 2(12), pp. 4023–4033.

[24] Wei, Y., Vats, G. and Noheda, B., 2022. Synaptic behaviour in ferroelectric epitaxial rhombohedral Hf0.5Zr0.5O2 thin films. *Neuromorphic Computing and Engineering*, 2(4), p. 044007.

[25] Mulaosmanovic, H., Ocker, J., Müller, S., Noack, M., Müller, J., Polakowski, P., Mikolajick, T. and Slesazeck, S., 2017, June. Novel ferroelectric FET based synapse for neuromorphic systems. In *2017 Symposium on VLSI Technology* (pp. T176–T177). IEEE.

[26] Majumdar, S., Tan, H., Qin, Q. H. and van Dijken, S., 2019. Energy-efficient organic ferroelectric tunnel junction memristors for neuromorphic computing. *Advanced Electronic Materials*, 5(3), p. 1800795.

[27] Thakoor, S., Moopenn, A., Daud, T. and Thakoor, A. P., 1990. Solid-state thin-film memistor for electronic neural networks. *Journal of Applied Physics*, 67(6), pp. 3132–3135.

[28] Yang, J. T., Ge, C., Du, J. Y., Huang, H. Y., He, M., Wang, C., Lu, H. B., Yang, G. Z. and Jin, K. J., 2018. Artificial synapses emulated by an electrolyte-gated tungsten-oxide transistor. *Advanced Materials*, 30(34), p. 1801548.

[29] Yang, Z., Zhou, Y. and Ramanathan, S., 2012. Studies on room-temperature electric-field effect in ionic-liquid gated VO_2 three-terminal devices. *Journal of Applied Physics*, 111(1), pp. 014506-1–014506-5.

[30] Zhou, Y. and Ramanathan, S., 2012. Relaxation dynamics of ionic liquid – VO_2 interfaces and influence in electric double-layer transistors. *Journal of Applied Physics*, 111(8), pp. 084508-1–084508-7.

[31] Oh, C., Kim, I., Park, J., Park, Y., Choi, M. and Son, J., 2021. Deep proton insertion assisted by oxygen vacancies for long-term memory in VO_2 synaptic transistor. *Advanced Electronic Materials*, 7(2), p. 2000802.

[32] Zhang, H. T., Park, T. J., Zaluzhnyy, I. A., Wang, Q., Wadekar, S. N., Manna, S., Andrawis, R., Sprau, P. O., Sun, Y., Zhang, Z. and Huang, C., 2020. Perovskite neural trees. *Nature Communications*, 11(1), p. 2245.

[33] Shi, J., Ha, S. D., Zhou, Y., Schoofs, F. and Ramanathan, S., 2013. A correlated nickelate synaptic transistor. *Nature Communications*, 4(1), p. 2676.

[34] Ha, S. D., Shi, J., Meroz, Y., Mahadevan, L. and Ramanathan, S., 2014. Neuromimetic circuits with synaptic devices based on strongly correlated electron systems. *Physical Review Applied*, 2(6), p. 064003.

[35] Du, H., Lin, X., Xu, Z. and Chu, D., 2015. Electric double-layer transistors: A review of recent progress. *Journal of Materials Science*, 50, pp. 5641–5673.

[36] Maass, W. and Zador, A., 1997. Dynamic stochastic synapses as computational units. *Advances in Neural Information Processing Systems*, 10, pp. 194–200.

[37] Neftci, E. O., Pedroni, B. U., Joshi, S., Al-Shedivat, M. and Cauwenberghs, G., 2016. Stochastic synapses enable efficient brain-inspired learning machines. *Frontiers in Neuroscience*, 10, p. 241.

[38] Erokhin, V., 2022. *Fundamentals of organic neuromorphic systems*. Cham: Springer International Publishing.

[39] Feofilova, E. P. and Mysyakina, I. S., 2016. Lignin: Chemical structure, biodegradation, and practical application (a review). *Applied Biochemistry and Microbiology*, 52, pp. 573–581.

[40] Park, Y. and Lee, J. S., 2017. Artificial synapses with short- and long-term memory for spiking neural networks based on renewable materials. *ACS Nano*, *11*(9), pp. 8962–8969.

[41] Jang, B. C., Kim, S., Yang, S. Y., Park, J., Cha, J. H., Oh, J., Choi, J., Im, S. G., Dravid, V. P. and Choi, S. Y., 2019. Polymer analog memristive synapse with atomic-scale conductive filament for flexible neuromorphic computing system. *Nano Letters*, *19*(2), pp. 839–849.

[42] Ren, Y., Yang, J. Q., Zhou, L., Mao, J. Y., Zhang, S. R., Zhou, Y. and Han, S. T., 2018. Gate-tunable synaptic plasticity through controlled polarity of charge trapping in fullerene composites. *Advanced Functional Materials*, *28*(50), p. 1805599.

[43] Gkoupidenis, P., Schaefer, N., Garlan, B. and Malliaras, G. G., 2015. Neuromorphic functions in PEDOT: PSS organic electrochemical transistors. *Advanced Materials*, *44*(27), pp. 7176–7180.

[44] van De Burgt, Y., Melianas, A., Keene, S. T., Malliaras, G. and Salleo, A., 2018. Organic electronics for neuromorphic computing. *Nature Electronics*, *1*(7), pp. 386–397.

[45] Lee, Y., Park, H. L., Kim, Y. and Lee, T. W., 2021. Organic electronic synapses with low energy consumption. *Joule*, *5*(4), pp. 794–810.

[46] Pecqueur, S., Vuillaume, D. and Alibart, F., 2018. Perspective: Organic electronic materials and devices for neuromorphic engineering. *Journal of Applied Physics*, *124*(15), pp. 151902-1–151902-11.

[47] Lee, G., Baek, J. H., Ren, F., Pearton, S. J., Lee, G. H. and Kim, J., 2021. Artificial neuron and synapse devices based on 2D materials. *Small*, *17*(20), p. 2100640.

[48] Shi, Y., Liang, X., Yuan, B., Chen, V., Li, H., Hui, F., Yu, Z., Yuan, F., Pop, E., Wong, H. S. P. and Lanza, M., 2018. Electronic synapses made of layered two-dimensional materials. *Nature Electronics*, *1*(8), pp. 458–465.

[49] Krishnaprasad, A., Dev, D., Han, S. S., Shen, Y., Chung, H. S., Bae, T. S., Yoo, C., Jung, Y., Lanza, M. and Roy, T., 2022. MoS_2 synapses with ultra-low variability and their implementation in Boolean logic. *ACS Nano*, *16*(2), pp. 2866–2876.

[50] Chen, Y., Zhou, Y., Zhuge, F., Tian, B., Yan, M., Li, Y., He, Y. and Miao, X. S., 2019. Graphene–ferroelectric transistors as complementary synapses for supervised learning in spiking neural network. *npj 2D Materials and Applications*, *3*(1), p. 31.

[51] Li, B., Li, S., Wang, H., Chen, L., Liu, L., Feng, X., Li, Y., Chen, J., Gong, X. and Ang, K. W., 2020. An electronic synapse based on 2D ferroelectric $CuInP_2S_6$. *Advanced Electronic Materials*, *6*(12), p. 2000760.

[52] Zhu, X., Li, D., Liang, X. and Lu, W. D., 2019. Ionic modulation and ionic coupling effects in MoS_2 devices for neuromorphic computing. *Nature Materials*, *18*(2), pp. 141–148.

[53] Wang, Y., Yang, Y., He, Z., Zhu, H., Chen, L., Sun, Q. and Zhang, D. W., 2020. Laterally coupled 2D MoS_2 synaptic transistor with ion gating. *IEEE Electron Device Letters*, *41*(9), pp. 1424–1427.

[54] Sun, L., Zhang, Y., Hwang, G., Jiang, J., Kim, D., Eshete, Y. A., Zhao, R. and Yang, H., 2018. Synaptic computation enabled by joule heating of single-layered semiconductors for sound localization. *Nano Letters*, *18*(5), pp. 3229–3234.

[55] Cao, G., Meng, P., Chen, J., Liu, H., Bian, R., Zhu, C., Liu, F. and Liu, Z., 2021. 2D material based synaptic devices for neuromorphic computing. *Advanced Functional Materials*, *31*(4), p. 2005443.

[56] Huh, W., Lee, D. and Lee, C. H., 2020. Memristors based on 2D materials as an artificial synapse for neuromorphic electronics. *Advanced Materials*, *32*(51), p. 2002092.

References

[57] Sengupta, A. and Roy, K., 2018. Neuromorphic computing enabled by physics of electron spins: Prospects and perspectives. *Applied Physics Express*, *11*(3), p. 030101.

[58] Fong, X., Kim, Y., Venkatesan, R., Choday, S. H., Raghunathan, A. and Roy, K., 2016. Spin-transfer torque memories: Devices, circuits, and systems. *Proceedings of the IEEE*, *104*(7), pp. 1449–1488.

[59] Amiri, P. K. and Wang, K. L., 2012. Voltage-controlled magnetic anisotropy in spintronic devices. *Spin*, *2*(3), p. 1240002.

[60] Hirsch, J. E., 1999. Spin hall effect. *Physical Review Letters*, *83*(9), p. 1834.

[61] Sengupta, A., Banerjee, A. and Roy, K., 2016. Hybrid spintronic-CMOS spiking neural network with on-chip learning: Devices, circuits, and systems. *Physical Review Applied*, *6*(6), p. 064003.

[62] Emori, S., Bauer, U., Ahn, S. M., Martinez, E. and Beach, G. S., 2013. Current-driven dynamics of chiral ferromagnetic domain walls. *Nature Materials*, *12*(7), pp. 611–616.

[63] Fukami, S., Zhang, C., DuttaGupta, S., Kurenkov, A. and Ohno, H., 2016. Magnetization switching by spin–orbit torque in an antiferromagnet–ferromagnet bilayer system. *Nature Materials*, *15*(5), pp. 535–541.

[64] Borders, W. A., Akima, H., Fukami, S., Moriya, S., Kurihara, S., Horio, Y., Sato, S. and Ohno, H., 2016. Analogue spin–orbit torque device for artificial-neural-network-based associative memory operation. *Applied Physics Express*, *10*(1), p. 013007.

[65] Sengupta, A. and Roy, K., 2016. Short-term plasticity and long-term potentiation in magnetic tunnel junctions: Towards volatile synapses. *Physical Review Applied*, *5*(2), p. 024012.

[66] Song, K. M., Jeong, J. S., Pan, B., Zhang, X., Xia, J., Cha, S., Park, T. E., Kim, K., Finizio, S., Raabe, J. and Chang, J., 2020. Skyrmion-based artificial synapses for neuromorphic computing. *Nature Electronics*, *3*(3), pp. 148–155.

[67] Sengupta, A. and Roy, K., 2017. Encoding neural and synaptic functionalities in electron spin: A pathway to efficient neuromorphic computing. *Applied Physics Reviews*, *4*(4), pp. 041105-1–041105-23.

[68] Kurenkov, A., Fukami, S. and Ohno, H., 2020. Neuromorphic computing with antiferromagnetic spintronics. *Journal of Applied Physics*, *128*(1), pp. 010902-1–010902-12.

[69] Goodwill, J. M., Prasad, N., Hoskins, B. D., Daniels, M. W., Madhavan, A., Wan, L., Santos, T. S., Tran, M., Katine, J. A., Braganca, P. M. and Stiles, M. D., 2022. Implementation of a binary neural network on a passive array of magnetic tunnel junctions. *Physical Review Applied*, *18*(1), p. 014039.

[70] Borders, W. A., Pervaiz, A. Z., Fukami, S., Camsari, K. Y., Ohno, H. and Datta, S., 2019. Integer factorization using stochastic magnetic tunnel junctions. *Nature*, *573*(7774), pp. 390–393.

[71] Youngblood, N., Ríos Ocampo, C. A., Pernice, W. H. and Bhaskaran, H., 2023. Integrated optical memristors. *Nature Photonics*, *17*(7), pp. 561–572.

[72] Vandoorne, K., Mechet, P., Van Vaerenbergh, T., Fiers, M., Morthier, G., Verstraeten, D., Schrauwen, B., Dambre, J. and Bienstman, P., 2014. Experimental demonstration of reservoir computing on a silicon photonics chip. *Nature Communications*, *5*(1), p. 3541.

[73] Tait, A. N., Nahmias, M. A., Shastri, B. J. and Prucnal, P. R., 2014. Broadcast and weight: An integrated network for scalable photonic spike processing. *Journal of Lightwave Technology*, *32*(21), pp. 3427–3439.

[74] Cheng, Z., Ríos, C., Pernice, W. H., Wright, C. D. and Bhaskaran, H., 2017. On-chip photonic synapse. *Science Advances*, *3*(9), p. e1700160.
[75] Ríos, C., Youngblood, N., Cheng, Z., Le Gallo, M., Pernice, W. H., Wright, C. D., Sebastian, A. and Bhaskaran, H., 2019. In-memory computing on a photonic platform. *Science Advances*, *5*(2), p. eaau5759.
[76] Stegmaier, M., Ríos, C., Bhaskaran, H., Wright, C. D. and Pernice, W. H., 2017. Nonvolatile all-optical 1 × 2 switch for chipscale photonic networks. *Advanced Optical Materials*, *5*(1), p. 1600346.
[77] Chakraborty, I., Saha, G. and Roy, K., 2019. Photonic in-memory computing primitive for spiking neural networks using phase-change materials. *Physical Review Applied*, *11*(1), p. 014063.

4
Artificial Neurons

4.1 Fundamental Principles of Neuron Design and Neuronal Action

There are a large variety of spiking neuron models and their specific usage may rely on the application being considered. Let us first consider a basic computational model for a spiking neuron based on leaky-integrate and fire dynamics [1]. Considering the input synaptic current to the neuron to be represented by a set of spikes at time instants t_f, the neuron's membrane potential dynamics is described by Eq. (4.1):

$$\tau \frac{dV_{mem}}{dt} = -V_{mem} + \sum_f \delta(t - t_f), \tag{4.1}$$

where V_{mem} represents the membrane potential and τ represents the membrane time constant. Once the membrane potential reaches a particular threshold V_{th}, the neuron generates an outgoing spike and the membrane potential gets reset to zero. In algorithmic implementations, this spiking behavior results in a discontinuity which translates to difficulties in neuron activation gradient calculation. Therefore, many algorithm formulations consider a "reset by subtraction" principle wherein the neuron's membrane potential is subtracted by V_{th} upon spiking. The neuron's membrane potential is held constant at the reset potential for a time duration, referred to as the refractory period. Usage of refractory period in SNN algorithm is also heavily dependent on the application as well as the training method. Another bio-realistic neuron dynamics considered in various works is a spike frequency adaptation mechanism referred to as homeostasis [2]. This causes the neuron's firing threshold to increase every time it fires. Many computational models also consider using an adaptive leak parameter that increases upon neuron firing. Including homeostasis effects, Eq. (4.1) can be rewritten as:

$$\tau \frac{dV_{mem}}{dt} = -V_{mem}(1 + L) + \sum_f \delta(t - t_f), \tag{4.2}$$

where the leak parameter L increases every time the neuron fires, else it decays exponentially.

Digital CMOS implementations of spiking neurons (for instance, the model outlined in Eq. (4.1)) usually consider a timestep discretized version of the differential equations governing the membrane potential dynamics along with a comparator to implement the thresholding operation. A timestep discretized implementation of the leaky-integrate and fire membrane potential dynamics can be given as:

$$V_{mem}(n+1) = \left(1 - \frac{1}{\tau}\right) V_{mem}(n) + \frac{S(n)}{\tau}, \qquad (4.3)$$

where n represents the timestep and $S(n)$ represents spike magnitude in case there is a spiking event at the n-th timestep. It is therefore clear that a digital CMOS implementation of Eq. (4.3) will consist of a multi-bit adder circuit along with other control circuitry such as multiplexers and comparators, thereby consuming significant hardware overhead.

An alternate route to efficient hardware implementation can be found in the domain of analog CMOS implementation [3]. These approaches stem back to the early works in the field of neuromorphic computing where circuits built using CMOS transistors operating in the subthreshold region were shown to mimic a wide variety of neuronal temporal dynamics such as spike frequency adaptation and refractory dynamics among others. The main motivation behind such analog CMOS implementations of spiking neurons is the similarity of ion flow across neuron channels to the diffusion mechanism of carrier transport in transistors operating in the subthreshold regime. Interestingly, the time constants of temporal

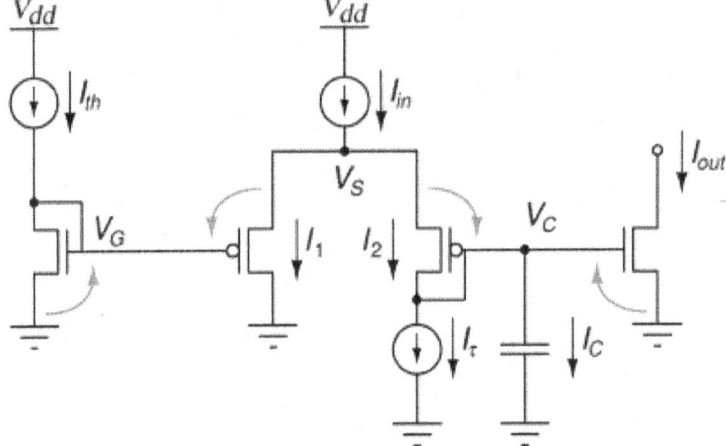

Figure 4.1 DPI circuit for emulating neuronal dynamics in analog CMOS circuits [3]. Adapted with permission from [3]. Copyright © 2014, IEEE.

dynamics associated with such circuits are of biological timescale (considering realistic capacitance sizes from a VLSI integration perspective as time constants scale linearly with the capacitance value) – thereby opening the possibility for using such circuit implementations for brain simulations or interfacing with bio-signals for biomedical signal processing.

Let us consider the differential-pair integrator (DPI) circuit [3] shown in Figure 4.1. Let us assume that by design, the following constraints are ensured: $I_{in} \gg I_\tau$ and $I_{out} \gg I_{th}$. Considering the rightmost transistor to operate in subthreshold saturation, we can write the drain current as,

$$I_{out} = I_o e^{\frac{kV_c}{U_T}}, \tag{4.4}$$

where I_o is the saturation current, U_T is the thermal voltage, and k is the subthreshold slope factor. Further, Eqs. (4.5)–(4.7) can be noted from the circuit:

$$I_c = C \frac{dV_c}{dt} \tag{4.5}$$

$$I_{in} = I_1 + I_2 \tag{4.6}$$

$$I_2 = I_\tau + I_c. \tag{4.7}$$

Further, by using the translinear principle wherein the sum of voltages in a chain of transistors operating in the subthreshold region can be expressed as multiplication of currents, we can write,

$$I_{th} \cdot I_1 = I_2 \cdot I_{out}. \tag{4.8}$$

From Eqs. (4.6)–(4.8), we can write,

$$I_{th} \cdot (I_{in} - I_\tau - I_c) = (I_\tau + I_c) \cdot I_{out}. \tag{4.9}$$

Utilizing Eqs. (4.4) and (4.5), we can simplify,

$$I_c = C \cdot \frac{U_T}{kI_{out}} \frac{dI_{out}}{dt}. \tag{4.10}$$

Using Eqs. (4.9) and (4.10), we can derive the following relationship (considering $\frac{I_{th}}{I_{out}} \to 0$ in the left-hand side and neglecting $-I_{th}$ on the right-hand side of the equation):

$$\tau \left(1 + \frac{I_{th}}{I_{out}}\right) \frac{dI_{out}}{dt} + I_{out} = \frac{I_{th} \cdot I_{in}}{I_\tau} - I_{th},$$

$$\Rightarrow \tau \frac{dI_{out}}{dt} + I_{out} = \frac{I_{th} \cdot I_{in}}{I_\tau} \tag{4.11}$$

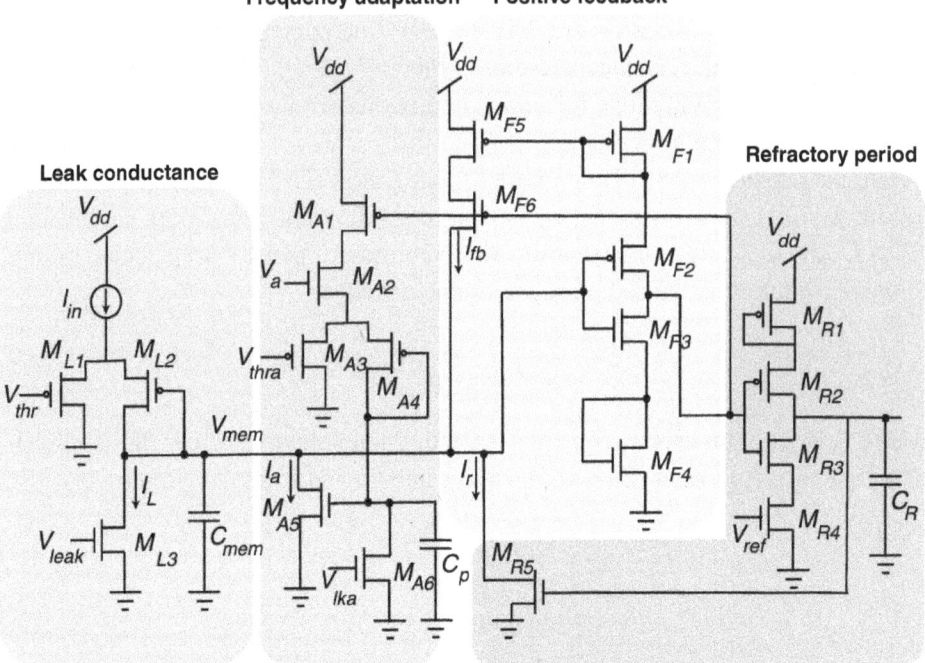

Figure 4.2 Subthreshold CMOS circuit with leak, homeostasis, and refractory properties [4]. Adapted with permission from [4]. © 2016 American Physical Society.

where $\tau = \frac{CU_T}{kI_r}$ is the time constant of the circuit. As can be clearly seen, the temporal dynamics closely resembles that of a leaky-integrate-fire neuron.

Additional functionalities of homeostasis and refractory period can be incorporated by considering the circuit as shown in Figure 4.2. Here, I_{in} is the resultant input current to the neuron integrated by the membrane capacitance C_{mem}. Transistor M_{L1} implements the leak functionality. V_{th} is the threshold voltage of the neuron. M_{T1} and M_{T2} form a source–follower circuit and triggers the inverter $M_{F1} - M_{F3}$ in order to generate V_{spike}. Homeostasis functionality is implemented by transistors $M_{A1} - M_{A4}$ which are used to adapt the spiking frequency through current I_a which increases as the neuron spiking frequency increases. Transistors $M_{F1} - M_{F5}$ are used to provide positive feedback using I_{fb} to increase V_{mem} quickly such that power consumption during state transition can be reduced. Transistors $M_{R1} - M_{R5}$ are used to implement refractory period as transistor M_{R5} absorbs current through I_r. While such designs provide a compact implementation of spiking neuron functionalities, they are difficult to design due to transistor size matching needs as well as scalability challenges due to variations. Figure 4.3 depicts the temporal evolution of membrane potential of the circuit when subjected to

4.2 Filamentary Neurons

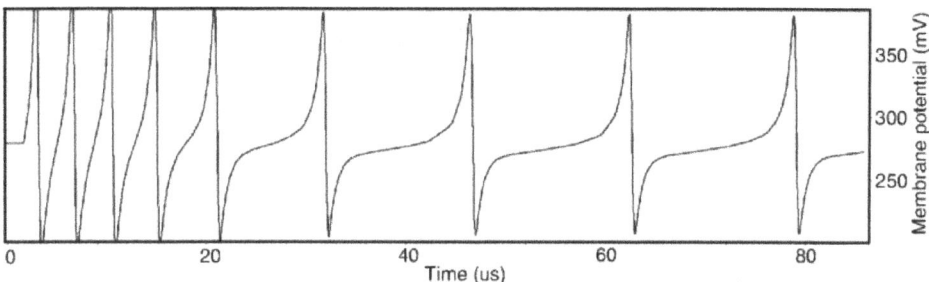

Figure 4.3 Time-domain response of the analog CMOS neuron in response to a constant input current demonstrating homeostasis functionality [4]. Adapted with permission from [4]. © 2016 American Physical Society.

a constant input current. As shown in the figure, no leak is observed due to the input of a constant current. Further, homeostasis in the form of spike frequency adaptation is also observed where the spiking frequency gradually decreases as the neuron spikes more and more. Energy consumption of such analog silicon neurons is usually in the ~pJ range for membrane capacitances of a few tens of fF [4].

In summary, analog CMOS circuit designs offer a promising pathway to brain-inspired computing, however, they are still not able to directly emulate the bio-realistic neuronal dynamics. Going forward, we will review various post-CMOS nanoelectronic device technologies that suggest routes toward direct mapping and study their relative merits and limitations to date.

4.2 Filamentary Neurons

Neuronal oscillations have been demonstrated in several semiconductor systems where local changes in conductance through formation of filaments or nanosized inclusions inside the channel leads to threshold switching. There are several types of filaments that can be formed, namely, metallic nanocluster or atomic pathways, point defect channels mediated by oxygen vacancies or local precipitates. We discuss representative examples of these various cases in this section. Note that the exact microscopic mechanisms and the associated thermodynamics (e.g., for defect formation and migration), kinetics (speed of resistance switching), and the energetics (voltages/fields) can vary among these different systems and should be considered carefully as they will dictate the energy barriers for state changes, operating frequency ranges, and reversibility or pinning (hysteresis). We discuss a few representative examples wherein the mechanism of oscillations can vary from vacancy point defect redistribution to metal dissolution.

InGaZnO$_4$ (IGZO) grown on TiN can demonstrate threshold switching with a top electrode made of Cu-Ta alloy. Figure 4.4 shows two types of I–V curves taken

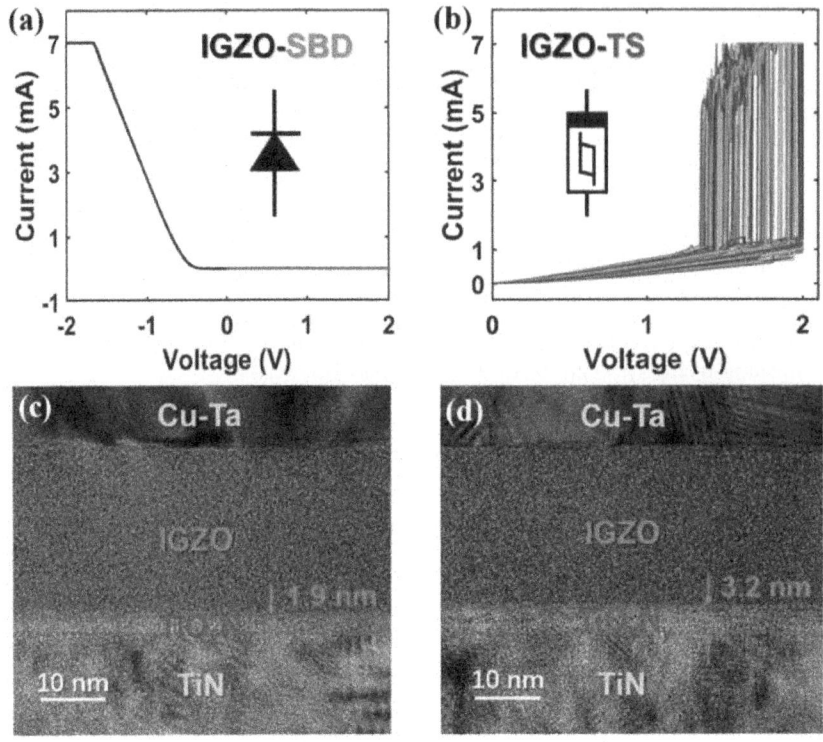

Figure 4.4 (a) Schottky diode behavior across the IGZO–TiN interface, (b) threshold switching is seen after the device undergoes forming step of 7.5 V, (c) cross-sectional TEM image of device before forming and (d) after forming. The reaction layer thickness at the IGZO–TiN interface is increased. The top electrode was Cu:Ta alloy of 35 nm thickness. The device size is 50×50 μm^2 [5]. Adapted with permission from [5], under the Creative Commons Attribution 4.0 license.

from a cross-bar type device [5]. Figure 4.4(a) shows Schottky diode characteristics before the forming step. The barrier is believed to occur at the IGZO–TiN interface. Once a forming voltage is applied, however the behavior changes dramatically. Figure 4.4(b) shows threshold switching, that is, a sharp rise in current at voltages approaching 2 V. This change is ascribed to the dissolution of copper ions from the top electrode and their drift into the oxide layer leading to conducting filaments which serve as soft breakdown paths. Figure 4.4(c) and (d) shows cross-sectional transmission electron micrographs before and after forming steps. A thin interfacial reaction layer of TiO_xN_y is observed whose thickness increased from 1.9 to 3.2 nm after the forming step. This could imply some of the oxygen from the IGZO layer was consumed to form this layer resulting in oxygen deficiency. These oxygen defects could result in the increase in leakage current through the layer as noted in Figure 4.4(b). Interestingly, threshold switching was not observed unless the top electrode comprised of Cu:Ta alloy. Pure copper as top electrode

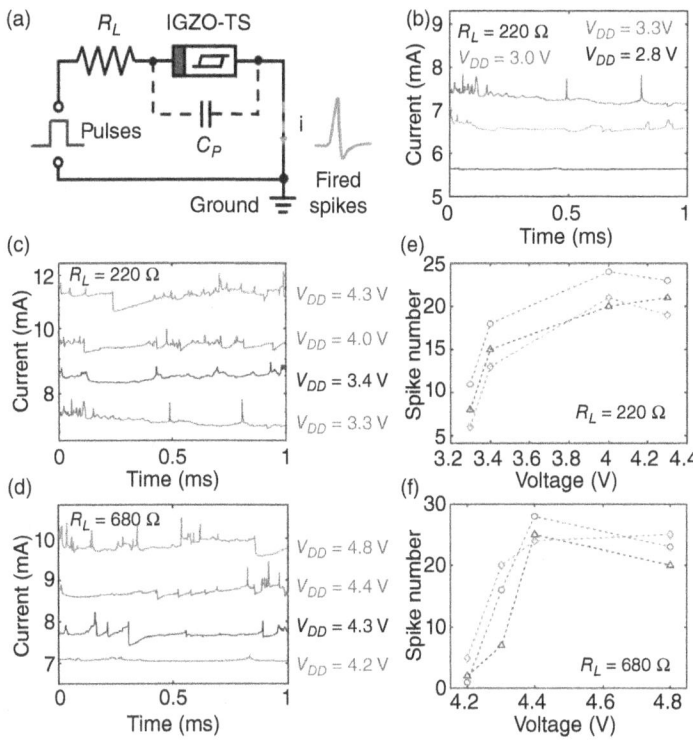

Figure 4.5 (a) Neuronal circuit configuration to achieve spiking behavior, (b) evolution of spiking as the voltage is increased from 2.8 V, (c) random nature of spiking for continuous increase in voltage, (d) spiking behavior while keeping the load resistance fixed at 680 Ohms, and (e and f) show dependence of number of spikes on voltage and load resistance [5]. Adapted with permission from [5], under the Creative Commons Attribution 4.0 license.

only resulted in bipolar behavior indicating the crucial role of the ability of electrode material being able to interact chemically with the switching layer. Hence, in this device concept, the electrode plays an important role in terms of chemical reactivity to enable threshold switching. The switching layer has an intrinsic capacitance which can be exploited in the neuronal circuit. Figure 4.5(a)–(f) shows the neuronal circuit configuration along with spiking patterns for different voltages applied. The charging and discharging of this internal capacitor enables the spiking oscillations. For 2.8 V, no spiking is observed while random spiking behavior was noted as the voltage increased gradually to 4.8 V. The number of spikes is also dependent on the load resistance. One reason for the observed stochasticity could be the random nature of Cu-In-O reaction products such as precipitates that are formed during the electrical cycling. The random nature of the spiking can in turn be utilized for solving certain types of neuromorphic computing problems, for example, path optimization and we will discuss this further in Chapter 7.

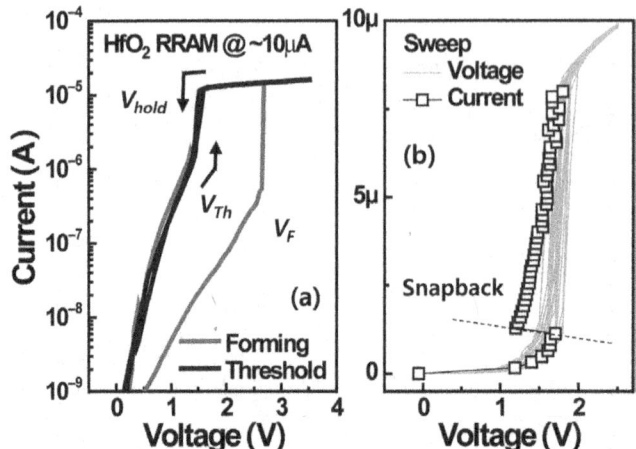

Figure 4.6 (a) Threshold switching characteristic of a hafnia device operated around 10 μA current level, (b) rapid increase in current levels below the 10 μA limit. Upon significant increase in current or voltage, the device can transition to a nonvolatile mode due to the robustness of the vacancy filaments. At higher voltages, it is possible to repurpose the device as a synapse [6]. Adapted with permission from [6]. © 2017 Elsevier B.V. All rights reserved.

Motion of oxygen vacancies (point defects) under voltage bias can be utilized to realize neuron function. Hafnium oxide is a widely studied electronic material in recent years due to its application as gate dielectric in highly scaled silicon transistors. Hafnia films can be grown by a variety of methods including both physical vapor deposition and chemical vapor deposition. By controlling the defect density in HfO_{2-x} films (where x represents non-stoichiometry in the oxygen sub-lattice) and further by careful choice of electric stimuli, it is possible to reorganize oxygen vacancy distribution across two electrodes to create a volatile (or threshold) switch. Figure 4.6(a) and (b) shows current–voltage sweeps taken from a 6-nm hafnia film sandwiched between TiN and Ti electrodes [6].

An abrupt increase in current is seen when the voltage reaches V_F in the first cycle leading to a low-resistance state (LRS). When the bias is removed, the device returns to a high-resistance state (HRS) indicating absence of a persistent memory effect. Subsequent sweeps of the voltage leads to similar threshold switching characteristic. Analysis of the current transport mechanism utilizing Poole-Frenkel model reveals the role of oxygen vacancies in determining the switching behavior. Creating a weak link of oxygen vacancies at low current levels results in abrupt switching, however the links dissolve when the stimulus is removed resulting in a threshold or volatile behavior. Hence, the device can in principle show self-oscillations when connected to a capacitor in parallel configuration. In other words, a threshold switch formed by filamentary point defects could find use as an

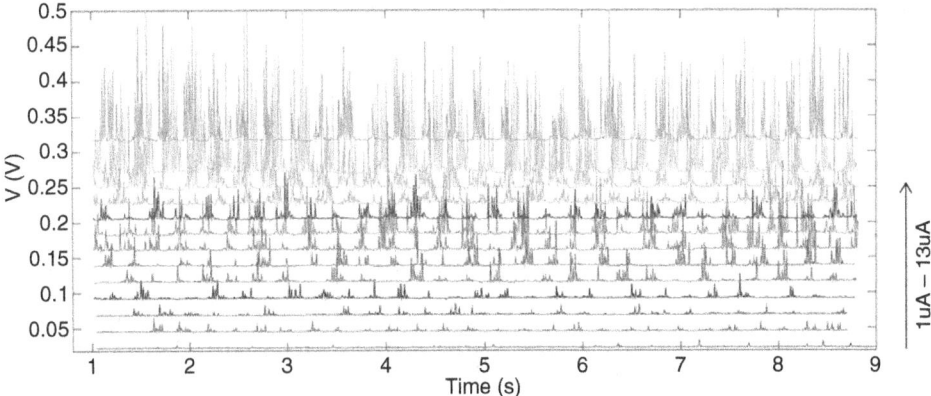

Figure 4.7 Voltage response of a silica-based two-terminal device as a function of constant current input. The firing frequency increases as current levels are increased [7]. Adapted with permission from [7], under the Creative Commons Attribution 4.0 license.

artificial neuron. In this scenario, it is possible to imagine a wide range of oxide or other materials that could be quite interesting for such applications. The key requirement here would be spontaneous recovery back to the initial state upon removal of the threshold electric stimulus that can reset the device.

Next, we consider another prototypical oxide, namely silica for exactly this purpose. Silica films of 37 nm thickness were sandwiched between 100 nm TiN electrodes for investigating resistance switching [7]. Both memory and threshold switching can be observed depending on the voltage or current levels applied to the device. Figure 4.7 shows how the voltage oscillations can occur for various constant currents being supplied to the device. Above a threshold current of the order 3 µA, the device shows spiking behavior. Additionally, the firing frequency increases with current in an analogous manner to biological neurons. A simplified version of the Hodgkin-Huxley model using a variable resistor in parallel to a capacitor was able to predict voltage instability in the case of memory switching while it is still a challenge for the case of volatile switching. At least for the case of silica, memory switching based voltage spiking was more stable than threshold switching. The details of SET process, currents needed to rupture filaments require further research as different mechanisms are responsible for these distinct steps necessary for the operation of a device in the nonlinear current transport regime. This is generally true for almost all filamentary switches due to the stochastic nature of filament formation and the abruptness in conductance jumps when the filaments reach percolation thresholds for soft breakdown.

While formation of localized filamentary pathways from point defect aggregation results in threshold switching, non-stoichiometric materials with mesoscale

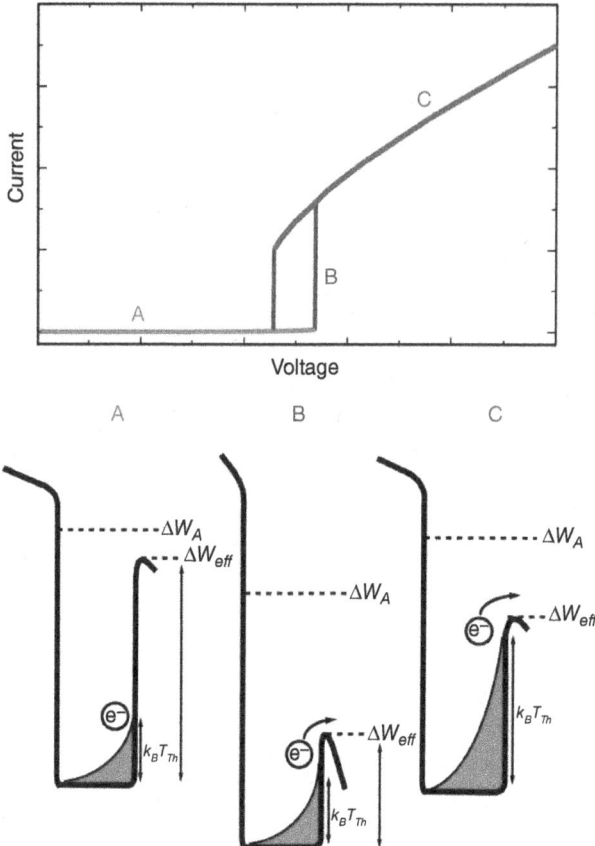

Figure 4.8 The top figure shows typical hysteretic I–V curve for a threshold switch. The regions marked A, B, and C correspond to HRS, abrupt jump in current at a threshold voltage and conducting state, respectively. ΔW_A is the energy barrier for charge carriers, while ΔW_{eff} is the effective barrier due to field-induced lowering, k and T_{th} refer to Boltzmann constant and temperature respectively. The simple band pictures for A, B, and C represent regions where the thermal energy is insufficient to excite carriers; thermal energy is comparable to field-driven barrier lowering and conducting state where local temperature is large enough to provide sufficient thermal activation, respectively [8]. Adapted with permission from [8]. © 2016 WILEY-VCH Verlag GmbH & Co. KGaA, Weinheim.

defect distribution can also display abrupt resistance jumps. Thermal runaway due to feedback from Joule heating has been proposed as a viable path to design threshold switches. A schematic of this process is illustrated in Figure 4.8. The Poole–Frenkel (PF) conduction mechanism enables effective reduction in barrier for electron flow across a dielectric approaching a critical field value. This results in self-acceleration and feedback due to Joule heating. The conducting state (region C in Figure 4.8) is reached when the thermal energy is sufficient to excite charge carriers over the dielectric–electrode energy barrier and is manifested as an abrupt jump in the current. By carefully engineering series resistors in the circuit, it is possible to limit the net current flow through the device and avoid irreversible breakdown. Interestingly, the origin of PF conduction could be material specific: in some cases arising from trap states due to oxygen defects, or simply polaronic hopping conduction. Different materials such as NbO_x and TaO_x have been reported as promising candidates to date. Note that such defective layers are often naturally formed on polycrystalline metal electrodes such as TiN or Pt during oxide film deposition since formation of stoichiometric line compounds (e.g., NbO_2) requires lattice matching and extremely tight window of oxygen partial pressure. Therefore, for vertical crossbar arrays, it is relatively straightforward to fabricate sub-stoichiometric layers on metal electrodes. Measurements of threshold voltage as function of ambient temperature for conducting state onset is a robust way to investigate the effect of Joule heating. Alternatively, materials prepared with well-controlled oxygen non-stoichiometry can be studied by spectroscopic methods to estimate sub-band gap trap energy distribution. With this knowledge, it is possible to correlate electrical switching thresholds with temperature since carrier excitation across an energy gap is a central feature of this mechanism. Since defect-mediated high field transport is possible in a broad range of insulators, this mechanism opens a wide selection of materials suited as building blocks of artificial neurons and selectors for resistive memory [8, 9].

4.3 Ferroelectric Neurons

The integrate and fire function noted in biological neurons can be elegantly emulated with ferroelectric (FE)-based artificial neurons via a few different approaches. First, the FE layer can be integrated into the gate stack of a transistor (referred to as ferroelectric field-effect transistor or FeFET). Alternatively, a ferroelectric tunnel junction (FTJ) can be utilized to realize artificial neuron function by polarization dependence of tunnel current. Ferroelectric materials can also be connected to CMOS circuits as separate layers to realize activation function at a higher circuit level integration. In this section, we discuss a few representative approaches starting with the FeFET. An FE film can be monolithically integrated onto a field-effect transistor in the gate stack. Figure 4.9 shows a schematic of such a device wherein an FE layer is fabricated on top of an interfacial dielectric layer (e.g., silica or similar insulator) deposited on a semiconductor [10]. The top electrode is then patterned on the FE layer. Although not essential, it is also possible to include a metal ground plane in the gate stack for improving the field uniformity in the FE.

Application of a gate voltage changes the orientation of the polarization domains in the FE. Depending on the direction of the polarization vector, the semiconductor

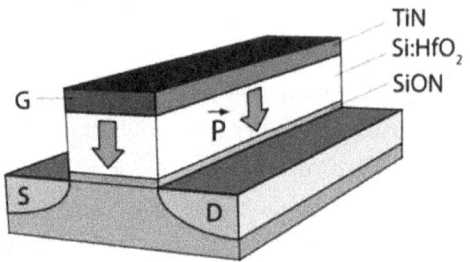

Figure 4.9 Schematic of an FeFET. The FE layer is interspersed between a dielectric layer and the metal gate. The arrows represent polarization direction of domains in the FE layer. Depending on the relative fraction of oriented domains along a certain direction, the source–drain current of the transistor will be affected due to electrostatic screening of charge carriers in the channel. The dielectric is a non-switching layer primarily to aid uniform deposition of the FE film and also minimize any leakage currents due to the smaller band gap of FE film [10]. Adapted with permission from [10].

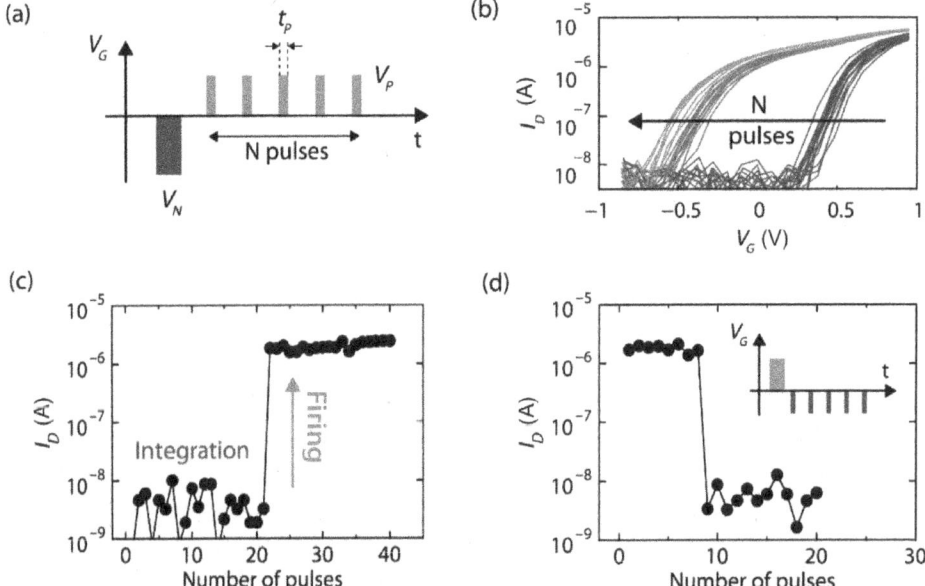

Figure 4.10 Accumulative switching of channel conductance in an FeFET device. The channel width is 80 nm and length is 30 nm. Upon application of successive voltage pulses shown in (a), the threshold voltage sharply shifts from high VT to low VT (from right to left as shown in (b)). In (c), the drain current is plotted with pulse number wherein the firing happens for pulse #21. In (d), the reverse ON to OFF transition is shown with reverse voltage polarity [10]. Adapted with permission from [10].

channel accumulates or depletes carriers in addition to the gate voltage effect. Hence, the threshold voltage for channel resistance switching depends on the polarization. The polarization state of the FE is nondestructively utilized for the bistable memory function in this device architecture by monitoring the drain current [11]. Note that the interfacial dielectric layer is not essential but often present due to fabrication process such as oxidizing conditions in the deposition environment that results in formation of a thin oxide layer on silicon. Under these scenarios, it is better to deposit a high-quality dielectric in a controlled manner and then deposit the FE layer instead of an uncontrolled dielectric layer formed by native oxidation. This is particularly important for non-silicon semiconductor channels where the native oxide could be discontinuous, contain large trap density or display large leakage characteristics. The interfacial layer will of course dictate the effective voltage that is dropped across the FE layer; hence, it is important to engineer the relative thickness and dielectric constant for optimal threshold operation of the FeFET device.

For integrate and fire function, it is important to have an internal mechanism that can keep track of past stimuli. For this purpose, we may utilize the fraction of domains oriented along the field direction. By continuous application of electric stimulus, the relative population of domains oriented along the field direction can be modulated resulting in incremental response until a critical population is reached for the "fire" function. To incorporate the leak function, either leaky integrators can be connected to the transistor gate or a leaky FE can itself be implemented to simplify the circuit design. Figure 4.10 shows experimental data concerning accumulative switching in an FeFET comprised of Si-doped hafnia FE layer sandwiched between TiN gate electrode and SiON dielectric on silicon substrate. The FE layer comprising polycrystalline 4 mol% Si-doped hafnia is 10 nm thick and the SiON dielectric is 1.2 nm thick. An abrupt switching of channel conductance is seen in Figure 4.10(c) when the 21st pulse is applied corresponding to the pulse scheme in Figure 4.10(a). Figure 4.10(b) shows the drain current characteristics with gate voltage pulses. Figure 4.10(d) shows the reverse process when the voltage polarity is switched. The action potential of biological neurons is emulated by the sharp rise in drain current. The mechanism is the following: Upon application of voltage pulses, a critical number of domains are nucleated between the source and drain regions resulting in a conducting pathway for the charge carriers. Below the criticality, the inversion is simply insufficient to modulate the conductance in a significant way.

Note that the neuron must be reset after the firing operation to return to the initial state. This can be accomplished by applying a voltage bias of opposite polarity. Signal integration with both polarities offer the possibility of implementing excitatory and inhibitory stimuli with FeFETs. Leaky characteristics could be emulated partially by application of smaller reset pulses in between the accumulation pulses. Other approaches to enhance the leak function could be to alter the relative thickness of the FE and dielectric layers to increase the depolarizing field which can spontaneously destabilize the FE domain orientations.

The abovementioned example has illustrated the basic integrate-fire function of the neuron. The leak function was naturally absent due to the stability of the FE properties but could be artificially induced by weakening the accumulation process. Another approach to naturally incorporate the leak function is simply to work with a FE layer that has time-dependent decay of the polarization. This could be realized for instance by using a leaky FE, that is, leakage currents flowing through the FE layer that serve to destabilize the polarization. With such a system, a delicate balance between integration and leak dynamics can be achieved. In fact, the same parent un-doped compound as in the abovementioned example, namely, hafnia has been exploited for this purpose. Incorporating a 5-nm Zr-doped hafnia film in FeFET enabled leak-integrate-fire function [12]. This was accomplished by combination of increased leakage current arising from a partially crystallized film and

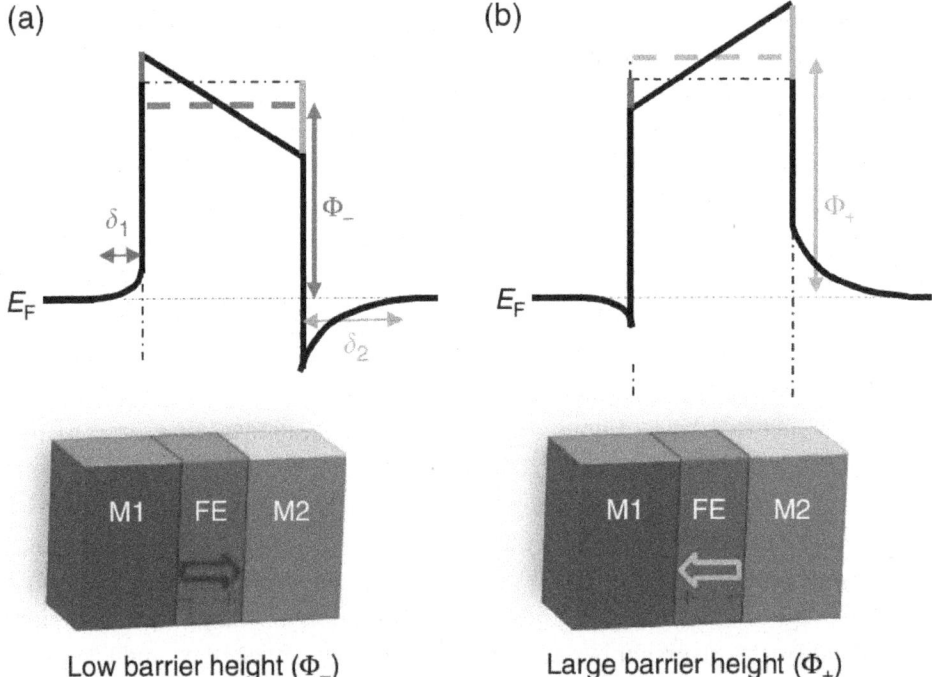

Figure 4.11 (a) Schematic on the top shows the band diagram for polarization vector pointing from left to right (bottom image), (b) while the top image shows the band diagram for polarization vector pointing from right to left (bottom image). M1 and M2 refer to the electrodes, and FE refers to the ferroelectric layer which acts as the tunnel barrier. Note that on an average, the barrier height for electron transport is larger when polarization is pointing from right to left than the other way around. This asymmetry results in a tunnel current resistance dependent on polarization [14]. Adapted with permission from [14]. Copyright © 2014, Springer Nature Limited.

the ultra-thin nature that enhanced the depolarization field. A reset pulse is then not required since the spontaneous decay of polarization will recover the initial state over the decay timescale. This can enable energy savings over multiple operations with the trade-off being reset dynamics determined by the internal physics of the device. In other words, careful engineering of CMOS-compatible ferroelectrics like orthorhombic hafnia can enable emulation of basic neuron functions.

Tunnel junctions incorporating ferroelectrics (FTJs) are also interesting to emulate neuronal function. CMOS circuits integrated with FTJs can amplify the junction current to realize activation. Partial polarization switching under appropriate voltage stimulus results in history-dependent tunnel current that can demonstrate accumulation characteristics, hence representing a tunable RC device [13]. The key physics is to understand the band structure of the tunnel junction and optimize the leakage current versus polarization-dependent switching current to maximize signal–noise ratio. Figure 4.11 shows the band diagram for an FTJ under different polarization conditions [14]. By careful control of the depletion region width at the electrode interfaces on either side of the FE layer, it is possible to obtain a tunnel resistance that is dependent on the relative orientation of the polarization vector. Such a junction forms the basis for a variety of solid-state devices for neural information processing and also for bistable memory. Hence, careful control of interfaces, thickness, and microstructure of the various dielectric, FE, and metal electrode layers are of utmost importance. Note that we discussed the use of FTJs as synapses in Chapter 3.

4.4 Insulator–Metal Transition-Based Neurons

Generating periodic voltage outputs for given electrical stimulus is an important characteristic of the biological neuron (i.e., action potential). Electrically driven insulator–metal transitions (IMTs) in correlated oxides are well suited for this purpose as threshold switches. Correlated oxide semiconductors represent a class of quantum materials whose ground-state properties cannot be predicted by traditional band theory. In this book, we are primarily concerned with their electrical switching characteristics. The electronic properties of such semiconductors are described elsewhere [15, 16]. One common characteristic of interest to us is the ability to drastically change their electrical conductivity upon changing temperature and/or applying an electrical bias. We begin by briefly discussing the thermally driven resistance switching which is closely related to the electrical analog.

Figure 4.12 shows a representative resistance–temperature plot for VO_2 [17]. VO_2 is a model system that has been studied for over six decades since the observation of IMT by Morin in 1959 and continues to attract interest. One important reason for this renewed interest is the fact that thin films can sustain reversible cycling across

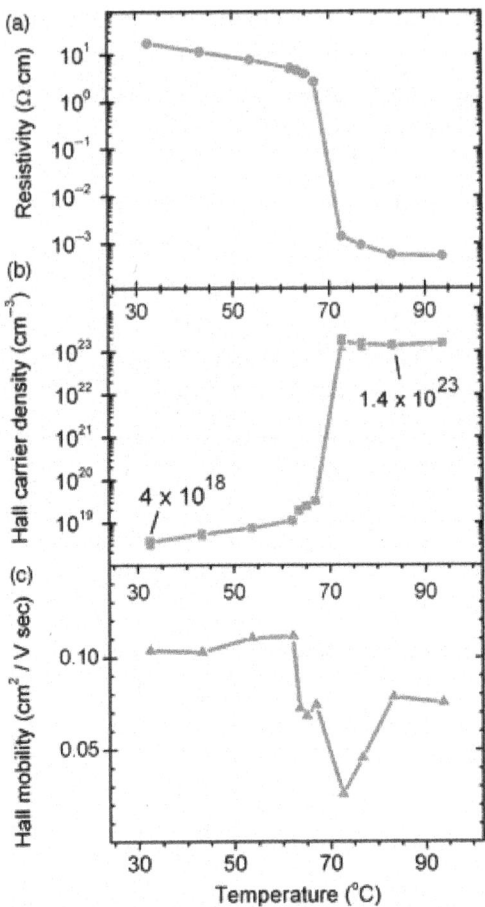

Figure 4.12 (a) Resistivity versus temperature for a VO_2 thin film showing a large drop in electrical resistivity as the temperature approaches 67°C. (b) The mechanism for resistivity drop is the increase in carrier density, (c) while the carrier mobility remains largely unchanged [17]. Adapted with permission from [17]. © 2009 American Physical Society.

the phase transition without failure for hundreds of millions of operations which was not simply possible in bulk crystals that would crack after just few heating-cooling cycles. It is clearly seen that there is a sharp drop in resistance of nearly four orders of magnitude near 67°C. Interestingly, the carrier density obtained from Hall measurements is primarily responsible for this change in conductivity while the mobility is nearly unchanged (only ~2x variation compared to nearly 10^4 variation in carrier density). The low mobility coupled with carrier density modulation being responsible for conductance transition represents a paradigm shift in semiconductor design. It is partially because of this unique property that polycrystalline correlated oxides also display excellent phase switching properties opening the path

4.4 Insulator–Metal Transition-Based Neurons

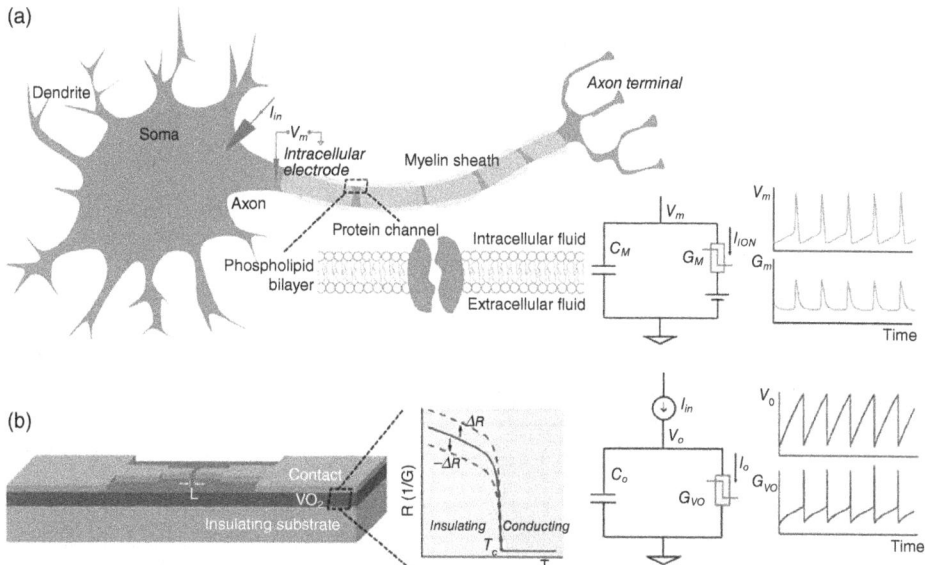

Figure 4.13 Biological versus IMT neuron. (a) The biological neuron can be considered as a parallel combination of membrane capacitance and conductance that fires an action potential at a critical threshold. (b) In an analogous manner, the IMT neuron undergoes conductance oscillations due to current injection when connected with a capacitor in parallel [18]. Adapted with permission from [18], under the Creative Commons Attribution 4.0 license.

to technological relevance. By varying the stoichiometry of the oxide films, it is possible to change the room temperature electrical conductivity (also referred to as ground-state resistance) as well as the transition temperature. If the transition temperature is reached by applying an electrical stimulus via Joule heating, then differing stoichiometry samples will have distinct threshold voltages for the conductance switching. Typically, as the oxygen vacancy concentration is increased, the material becomes more conducting and hence a lower voltage stimulus can drive the phase transition. This is a useful knob to tune the electrical properties, especially in a circuit if multiple neurons with differing thresholds are to be connected.

Next, the basic principle to design an artificial neuron using a VO_2 switch as an example is presented in Figure 4.13. The biological neuron can be described as a parallel combination of membrane capacitance and conductance in the framework of the Hodgkin–Huxley model. In an analogous manner, the IMT material undergoes a sharp conductance transition at a threshold voltage or current input due to Joule heating and can self-oscillate between the conducting and insulating states when connected to a capacitor in parallel [18]. This strikingly mimics the functional characteristics of the biological neuron while retaining a simple circuit configuration.

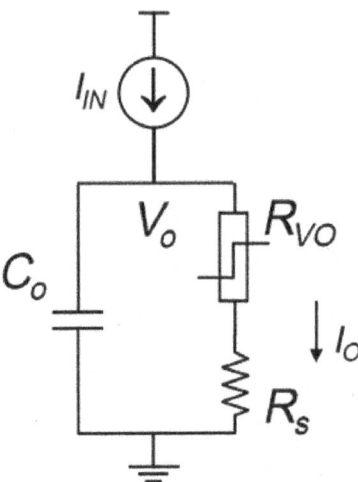

Figure 4.14 The IMT neuron circuit comprises of three elements, namely, a capacitor, IMT resistor, and a series resistor that acts as a sensing element. The current flow through the IMT material varies depending on the insulating versus conducting state [18]. Adapted with permission from [18], under the Creative Commons Attribution 4.0 license.

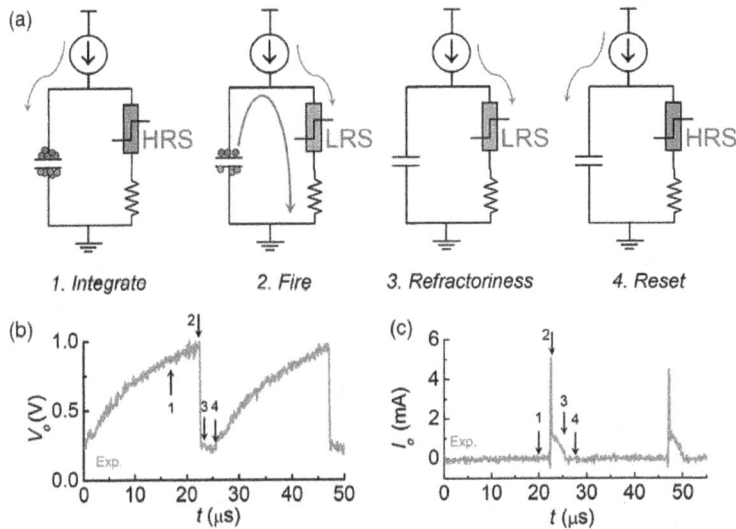

Figure 4.15 Various stages of neuronal firing and reset. In step 1, the capacitor is being charged as the VO_2 is in HRS. In step 2, the critical threshold voltage is reached to switch the VO_2 into LRS and the capacitor discharges resulting in a large current spike. In step 3, the VO_2 remains in LRS and drains excess charge to ground as it cools back to the insulating state. In step 4, the neuron resets and is ready to begin operation. In figure (b), the experimental output voltage is shown, and in (c), the experimental current output is presented for the four stages in (a) [18]. Adapted with permission from [18], under the Creative Commons Attribution 4.0 license.

Let us consider the working principle in more detail. A simple two-terminal device is presented first as shown in Figure 4.14. The VO_2 device undergoes an IMT when a sufficient current is injected. To emulate neuronal function, one can refer to the schematic in Figure 4.15. The material oscillates between conducting and insulating states resulting in current oscillations. In addition to the emulation of fundamental threshold spiking behavior, it is possible to realize more complex neuron characteristics including stochastic behavior by combining relaxation oscillators [19].

Other materials that undergo electrically driven phase transitions are also being considered for artificial neuron applications. Examples include NbO_2 that undergoes threshold switching in a similar manner to VO_2. Figures 4.16 and 4.17 show measured threshold switching characteristics and circuit configuration along with

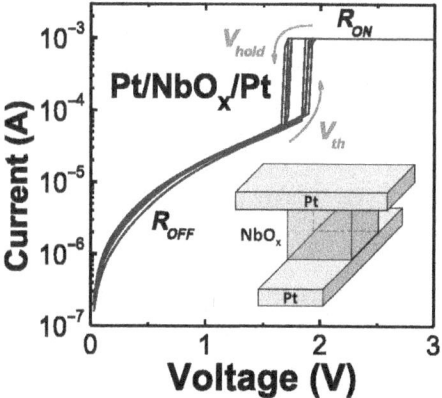

Figure 4.16 Experimentally measured current–voltage characteristics across an NbO_x device. The inset shows a schematic of the device. Refer to the text for a description of the threshold behavior [20]. Adapted with permission from [20]. 2017 © AIP Publishing.

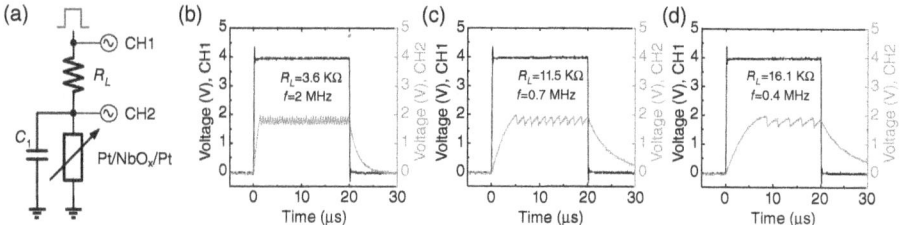

Figure 4.17 (a) Circuit diagram showing the IMT device connected to a load resistor. The graphs in (b)–(d) show oscillation characteristics as the load resistor R_L is varied. Clearly, the frequency of the oscillations depends on the series resistance value [20]. Adapted with permission from [20]. 2017 © AIP Publishing.

oscillation modes of a vertical Pt/NbO$_x$/Pt device [20]. The x subscript instead of the expected "2" value refers to the fact the oxide film is likely non-stoichiometric.

This is often the case when the oxide layer is grown on a non-lattice matched substrate. The threshold voltage in Figure 4.16 corresponds to the value on the x-axis at which there is an abrupt increase in the current flowing through the device, which is around 1.9 V in this case. When the voltage is swept back to 0, the current switches to the OFF state at near 1.6 V which is referred to as the hold voltage. It is worth noting that in this device there is no dependence on polarity of the voltage on the electrical characteristics, that is the device is nonpolar. The circuit to obtain oscillations is shown in Figure 4.17.

The IMT neuron node is connected to a series resistor that acts as a synapse. The resistance of R_L is chosen between the OFF and ON state resistance of the IMT device. The operational principle of this circuit is rather like the previous discussion on VO$_2$ based neurons. Note the oscillations have a triangular waveform since the charging timescale is larger than that of discharging. Another important

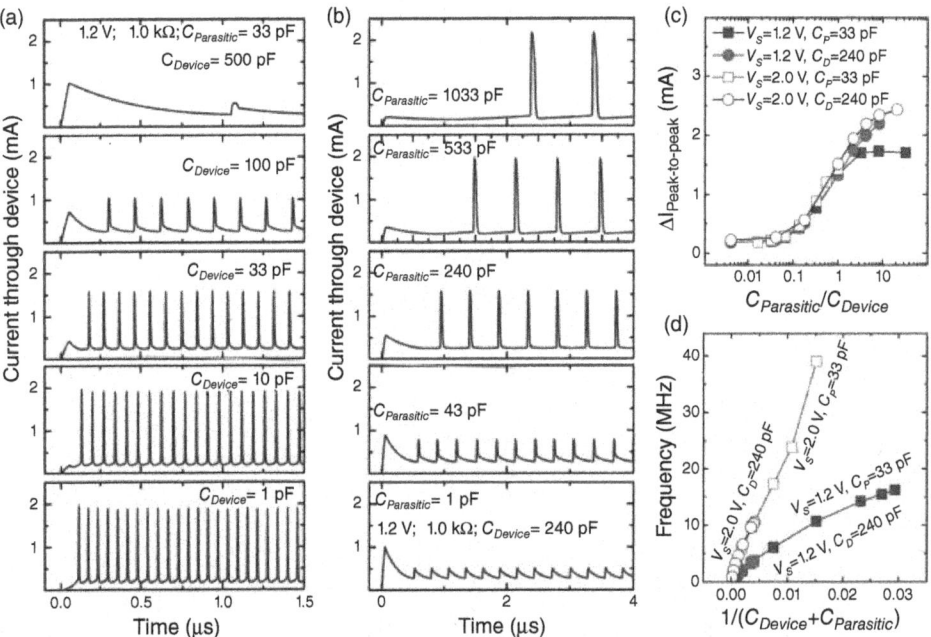

Figure 4.18 Characteristics of NbO$_x$ oscillators for various parasitic capacitances involved in the test circuit. (a) The parasitic capacitance is fixed at 33 pF, while the device capacitance is varied from 1 pF to 500 pF; (b) device capacitance is fixed at 240 pF, while the parasitic capacitance is varied from 1 pF to 1 nF; (c) variation of peak-to-peak amplitude of total device current with ratio of parasitic capacitance to that of device; and (d) frequency of oscillations depend inversely on the total capacitance in the circuit [21]. Adapted with permission from [21]. 2016 © AIP Publishing.

aspect is that the oscillation frequency limit is often constrained by the parasitic capacitances involved in the test setup and so the intrinsic timescales of relaxation could be significantly different from the measured values in the neuronal circuit configuration. One representative example of this scenario is shown in Figure 4.18 for the case of NbO_2 oscillators [21].

The oscillation period, current maxima, and frequency of oscillations closely depend on the parasitic capacitances and limit the operational envelope. It has also been proposed that the lifetime of carriers in the metallic phase can determine the relaxation kinetics back to the insulating state and hence the overall oscillation frequency in vanadium dioxide-based devices [22]. In this picture, the electronic properties of the carriers in the phases formed under electrical stimuli dictate the overall dynamics of the neuron behavior. This requires further research on how to control the carrier lifetimes and recombination dynamics in such phase transition-based systems.

For practical applications in arrays of neuromorphic devices, it is important to increase the ON/OFF ratio between the insulating and metallic states. Strategies to optimize the ratio while maintaining reliable device operation has been discussed in literature [23]. Due to the electrothermal coupling inherently present in such IMT devices, the local thermal dynamics is equally important as the electrical parameters of the circuit. This leads to important differences between CMOS-based neurons and IMT-based neurons in terms of parameters controllable by the circuit designer versus geometrical factors of the device [24]. The thermal sensitivity of the IMT neurons can also be exploited to design stochastic phase switches that may find use in homeostasis in networks [25].

The electronic phase transition can be induced by carrier doping in several oxide systems. Among the most prominent are the family of rare-earth perovskite nickelates where hydrogen doping of the parent compound has been shown to result in several orders of magnitude change in resistivity. The hydrogen resides as a proton in interstitial sites in the perovskite lattice while the electron is donated to the nickel e_g orbitals as verified from X-ray absorption spectroscopy. Electric fields can be utilized to drift the protons in the lattice. Since the position and concentration of the protons sensitively affect the electrical resistance of the nickelate channel, it is possible to emulate the action potential of neuron firing by pulsed electric fields in this material system. In addition, it is possible to emulate stochastic nature of neuronal action by controlling the strength of the electric pulse stimulus and the bias width [26]. A distinct feature of this approach is the nonvolatile nature of neurons in contrast to threshold switch-based neurons that decay back to original state when the stimulus is removed. This adds a level of flexibility to the designer for reset operation.

Besides the oxide-based devices discussed here, other classes of materials that undergo IMTs under critical stimulus can be exploited as leaky-integrate-fire neurons. One example is the lacunar spinel compound such as AM_4Q_8 where A could be Ga, Ge, and M could be V, Nb, and Mo, and Q is typically S or Se. The band gaps of such materials are of the order of 0.1–0.3 eV, and the insulating state can be collapsed into the conducting state by means of avalanche breakdown under strong electric fields approaching few kilovolts per centimeter [27]. Hence, electrically modulating conductance state by collapsing or opening a bandgap offers a powerful approach to design of artificial neurons. The electronic properties of Mott insulators are closely related to the impurity concentration. Carriers can get excited into the conduction band under strong electric fields resulting in breakdown of the insulating state. Hence, control of the ground state carrier density by introducing charged defects is an effective strategy for modifying threshold voltages of spiking artificial neurons. Processing methods used in traditional semiconductor processing such as ion irradiation can be adapted for phase transition systems to create disorder resulting in reduced threshold switches [28].

4.5 Organic Materials-Based Neurons

Organic materials offer intriguing prospects to design of artificial neurons from the perspective of bio-realistic action potential generation utilizing ion channels, in vivo interfacing with biological neuronal circuits and millisecond operational speeds. Further, the soft nature of the materials could be exploited for flexible electronics and monolithic integration with robotic exostructures to create sensory interfaces and haptics. There is a wealth of research activity in this field to realize artificial neural network hardware with organic systems. Organic semiconductors whose volumetric conductance can be modified by ionic intercalation are particularly suited for fabrication of electrochemical transistors. Further, if the doping can be n or p type depending on the material system, by careful combination of different devices, simple circuit building blocks can be designed to obtain the desired output characteristics. In this section, we will discuss a few representative efforts to demonstrate and characterize artificial neurons. To date, artificial neuron demonstration involves use of multiple organic semiconductor-based transistor devices interconnected with capacitors and resistors. The transistors act as the nonlinear elements while the resistor and capacitor in the circuit generate the spiking oscillatory outputs. We first discuss organic electrochemical transistors (OECTs) whose channel resistance is modulated by ionic doping/de-doping from an electrolyte gate. Figure 4.19 shows a schematic of biological neurons and the artificial analog using OECT circuit along with output curves [29]. The Axon-Hillock circuit comprises of an n-type and p-type transistor fabricated via screen printing and coating methods.

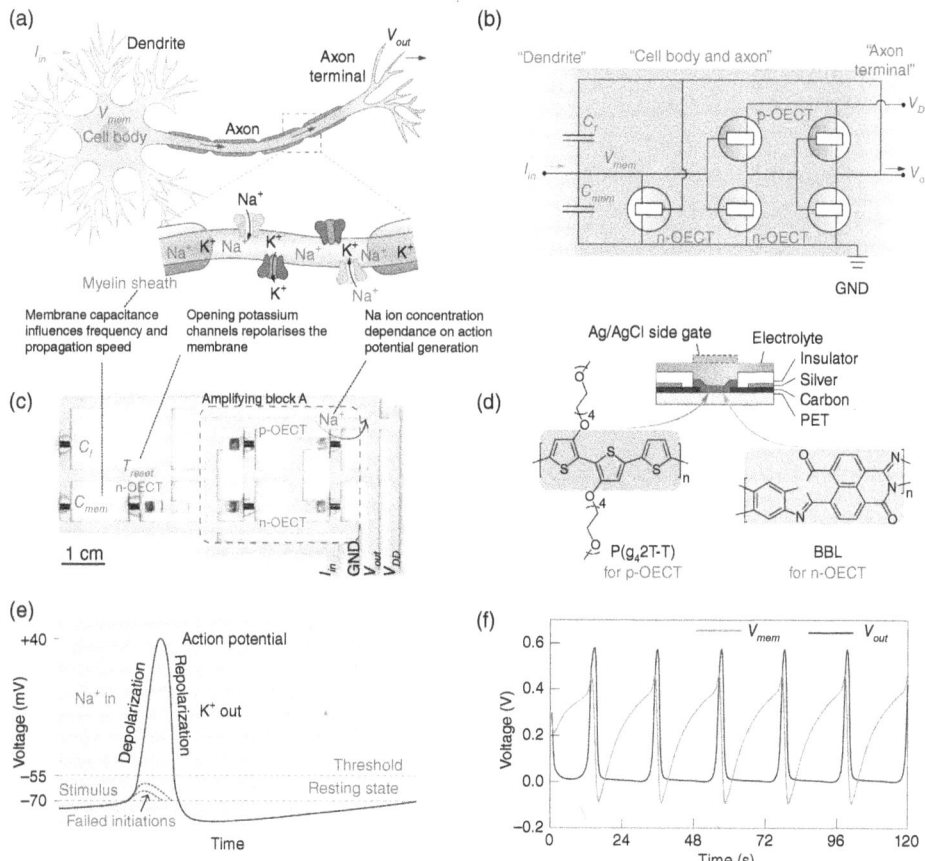

Figure 4.19 (a) Schematic of a biological neuron with analogy to the organic circuit counterpart. (b) Circuit building blocks to realize Axon-Hillock neuron. (c) Printed neuronal circuit for leaky-integrate-fire action. (d) Molecular structure of the organic materials used to fabricate the OECTs. (e) Action potential profiles in a biological cell and (f) experimental spiking characteristics of the OECTs. In this circuit, the thickness of the p- and n-type layers were optimized independently as 20 nm and 250 nm to obtain desired inverter characteristics, while the width and length were kept identical at 2,000 μm and 200 μm, respectively [29]. Adapted with permission from [29].

The devices are fabricated on a PET (polyethylene terephthalate) substrate with Ag/AgCl gates. The p-type semiconductor was glycolated polythiophene and the n-type semiconductor was poly(benzimidazobenzophenanthroline) and Ag/AgCl was the gate. The circuit was chosen to resemble the opening and closing of voltage-gated ion channels in biological neurons. When current is injected into the input terminal, the capacitor C_{mem} is charged thereby increasing V_{mem} gradually. Upon reaching a threshold, a voltage spike V_{out} is fired. The capacitor charges until the transistor threshold is reached and removal of input current during this process will result in decay of the capacitance, akin to failed neuronal spikes. Hence, threshold and subthreshold tuning of the signal strength is possible. Once the output voltage is high enough, the capacitor discharges through a resetting transistor T_{reset} in Figure 4.19(c). The V_{mem} can therefore reduce to the threshold voltage of the transistor and leads to a sharp drop in V_{out}. Hence, the entire process of generating an action potential can be restarted. The dynamics of firing can be tuned by the capacitor C_{mem} since a lower capacitance correspondingly decreases the charging time. The ionic currents generating spikes naturally resemble the biological neuron function at the cost of low-frequency operation due to the diffusional operation mechanism. At the same time, it is possible to further optimize the circuit parameters and device geometry to push the spiking frequency closer to that of the natural neuronal circuits.

Note that different combinations of organic transistors can be utilized for artificial spiking neuron demonstration. In another study, a nonlinear building block comprising of two p-type transistors were connected to RC circuits for oscillatory output [30]. The two transistors operated in depletion versus enhancement mode and were made from PEDOT-PSS and p(g2T-TT) semiconductors, respectively. By realizing negative differential resistance (NDR) with the two connected devices, it is possible to then construct the remaining circuit elements to create self-sustained oscillations. The use of liquid electrolytes opens several interesting connections to biology. First, the liquid is reminiscent of extracellular medium in the brain which sustains the ionic fluxes. The concentration of ions in the electrolyte dictates the ON–OFF characteristics of the transistor and thereby controls the spiking action. Clearly, this is analogous to delicate ionic equilibria in the brain that alters spike train frequencies. The low frequencies typically noted in ionic

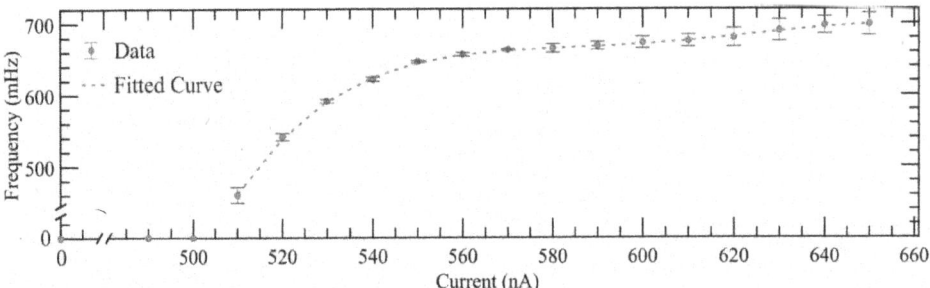

Figure 4.20 Axon-Hillock neuron circuit was fabricated as described in [32]. The plot shows dependence of spiking frequency as a function of input current. The frequency increases smoothly with the current injection eventually saturating. The neuron firing begins at a critical current value that is determined by the threshold voltage of the inverter transistors. The peak frequency is limited by the circuit parasitics due to the use of nonintegrated components in this case. Adapted with permission from [32]. 2020 © IOP Publishing.

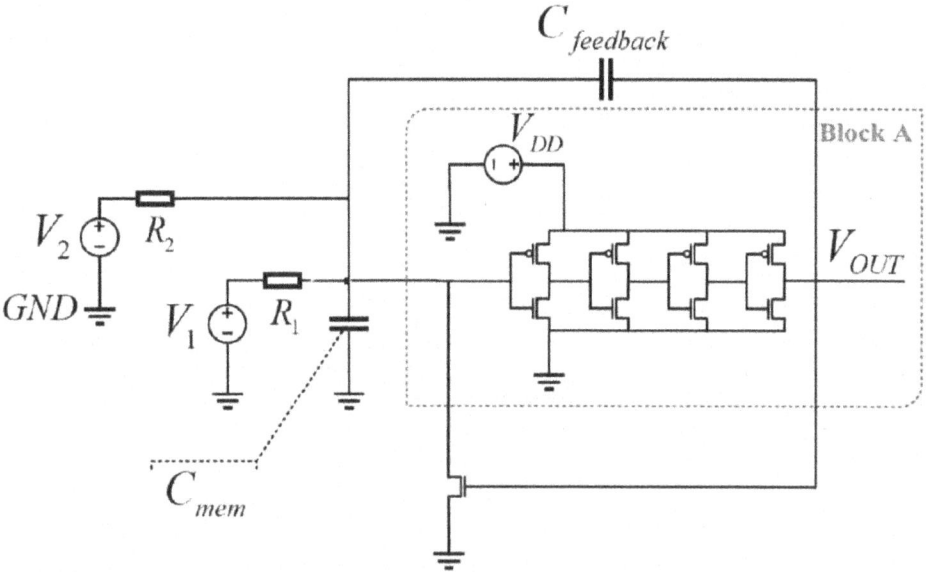

Figure 4.21 Circuit diagram of an Axon-Hillock neuron. A four-stage CMOS inverter is utilized here. A capacitor C_{mem} is connected to an amplifier block (labeled A). The output voltage spikes depend on the capacitance of C_{mem}. To emulate multiple synaptic inputs into the neuron, two voltage sources V_1 and V_2 are connected through resistances R_1 and R_2 to the capacitor node to demonstrate current summation effects [32]. Adapted with permission from [32]. 2020 © IOP Publishing.

or electrochemical transistors can match biological neural circuit speeds thereby opening possibilities for co-integration of brain matter with electronic circuits.

Integrate and fire neuron circuits using the p-type organic semiconductor DTBDT-C6 have been fabricated by printing on PEN substrates. An Axon-Hillock circuit comprising three transistors and three resistors were demonstrated. Parylene deposited by chemical vapor deposition was utilized as a gate dielectric [31]. Other organic semiconductors that have been utilized for demonstrating proof-of-principle A-H circuits include dinaphtho[2,3-b:2',3'-f]thieno[3,2-b]thiophene (DNTT) and N,N'-bis(n-octyl)-x:y,dicyanoperylene-3,4:9,10-bis(dicarboximide) (PDI8-CN2) as p- and n-type materials respectively [32]. Figure 4.20 shows frequency response of the neuron to input current.

The firing rate closely depends on the injected current and increasing the current can result in increasing frequency by nearly half an order of magnitude. Such circuits are also elegant for demonstrating Boolean summation. Multiple current sources (emulating synaptic inputs) can be integrated by the membrane capacitor that connects to the amplifier block as shown in Figure 4.21. While the strength of the individual current sources may not be sufficient to trigger an action

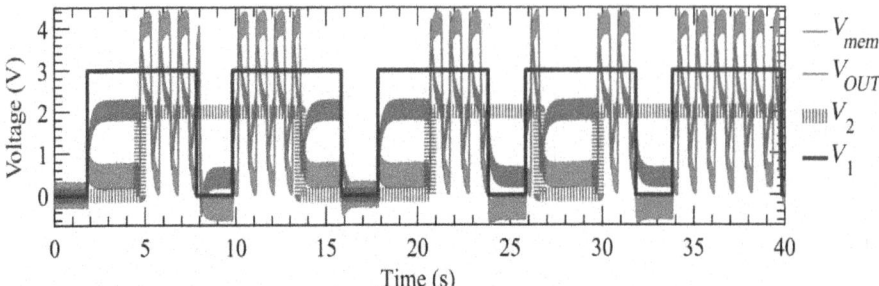

Figure 4.22 Boolean summation of currents into A-H neuron to demonstrate spiking only when threshold is reached. When the voltages V_1 and V_2 are paired together, the neuron spikes (see V_{OUT}). Else, the change in the membrane potential is below the threshold. The methodology can in principle be extended to a larger neural circuit where multiple synaptic inputs are summed at a neuron node [32]. Adapted with permission from [32]. 2020 © IOP Publishing.

potential, the combined currents could reach the necessary threshold. This is seen in Figure 4.22. Currents through R_1 and R_2 in Figure 4.21 are each insufficient to trigger an output voltage spike, however when they are presented together at the capacitor C_{mem}, summation of the currents results in spiking action [32].

Clearly, several organic semiconductors are interesting for demonstration of artificial neurons. Modulating channel conductance using classical semiconductor dopant action, that is, static accumulation or depletion of carriers at the channel–dielectric interface by a gate voltage or bias-driven incorporation or extraction of ionic dopants into the channel via mass exchange and diffusion are both feasible. However, factors to keep in mind include the differences in carrier mobility between p- and n-type semiconductor channels that require careful optimization of device geometry for fabrication of complementary inverters as well as increased fabrication complexity due to need for multiple materials to make p- and n-type channel devices. While many efforts to date include prototype demonstration using off-chip interconnections and discrete components connected through circuit boards, integrating all device elements onto a chip-compatible processing scheme will be essential in future for compact neuron function especially when they must be connected to several other neurons and synapses and external wires. Once the main active components are integrated monolithically, it will become possible to understand the critical factors that limit operational frequencies and power constraints for proper comparison to other technologies.

4.6 Two-Dimensional and Layered Material Neurons

Much work to date has focused on design of (leaky-)integrate-fire neurons utilizing volatile threshold switching in 2D or layered materials. Note here we are

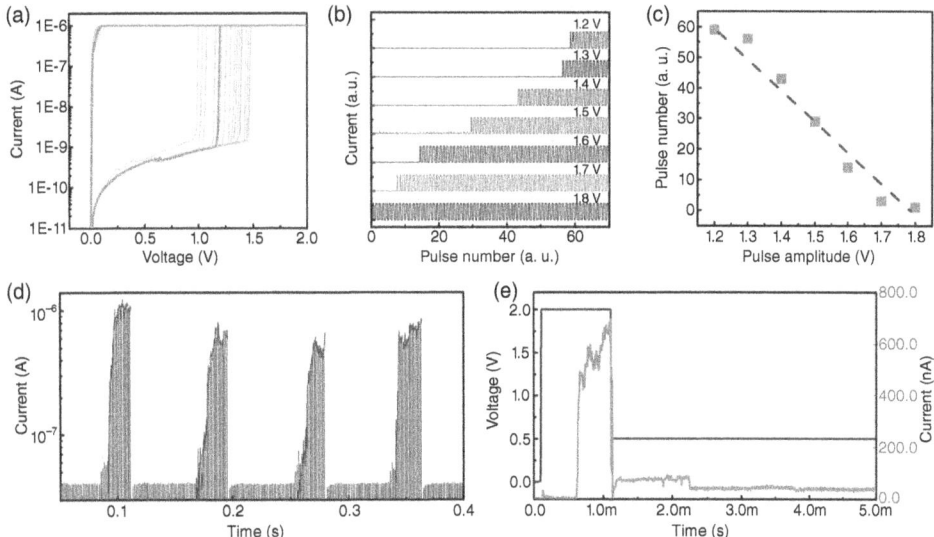

Figure 4.23 (a) Fifty current–voltage curves taken from monolayer MoS$_2$ showing abrupt jump in current at critical voltage near 1.2 V. (b) Under pulsed mode, the current increases for fewer pulse numbers indicating an integrate and fire capability. As an example, it takes over 50 pulses to initiate current jumps (firing action) for 1.2 V whereas firing initiates at under 20 pulses when the bias is increased to 1.6 V. (c) The relationship between pulse number to initiate firing versus pulse amplitude is shown with a clear inverse relationship. (d) Spontaneous recovery of the initial resting state after the voltage is removed. (e) Rise in current following stimulus pulsed bias of 2 V followed by reading the current at a lower voltage of 0.5 V. Reducing the voltage below the hold value results in resetting the device [33]. Adapted with permission from [33]. © 2020 WILEY-VCH Verlag GmbH & Co. KGaA, Weinheim.

referring to systems that can be fabricated as monolayers or few layered structures, and the devices demonstrated in this class are not necessarily from a single monolayer. This is because the field of artificial neurons using layered semiconductors is still in its infancy while it is much easier to demonstrate synaptic characteristics. The thickness dependence of volatile versus nonvolatile switching and the associated mechanisms creates the distinction between their use in neuronal (threshold) versus synaptic (memory) device applications.

Volatile switching has been realized in MoS$_2$ using silver ions as mobile species [33]. Monolayer MoS$_2$ single crystal in a planar device architecture with TiW and Ag electrodes can be used to emulate neuronal characteristics. The Ag electrode is intentionally chosen as silver ions are mobile under electric fields and loosely mimic the ion migration channels in biological neurons. A 500-nm channel length ensured volatile switching that enables the reset operation when the voltage is reduced below a threshold. In Figure 4.23(a), as the voltage increases to about 1.2 V, a sharp jump in current of the order of 10,000× is seen. Fifty sweeps are shown with the fifth sweep highlighted. Figure 4.23(b) and (c) shows how the current spikes increase when pulses of voltage are applied as opposed to continuous DC sweeps. While it takes over 50 pulses to initiate firing for 1.2 V across the electrodes, this value reduces to less than 10 for 1.7 V and trend is captured in Figure 4.23(c). Figure 4.23(d) and (e) shows the spiking behavior for voltage pulses above the threshold and spontaneous recovery upon removal of the bias. Combined, these results demonstrate proof-of-principle functionality of neurons using the integrate-fire function.

The mechanisms behind this behavior can be summarized as follows in a qualitative manner: Silver ions migrate across the MoS_2 channel under electric bias. At a critical bias, a conducting filament is formed resulting in the abrupt jump in current. Removal of the electrical stimulus dissolves the filament, while sequential pulse bias application retains some memory of the stimulus that resembles the integration function. In this study, the authors performed similar resistance switching studies on a silica control (replacing the MoS_2) and found no evidence for threshold switching. This led them to conclude the crucial role of the monolayer material with its superior mobility for silver ions (activation energy is around 0.14 eV in MoS_2) being responsible for the observed LIF characteristics. Additional details become important for reproducibility. For instance, if only one filament is formed during repeated switching cycles, then we can expect the cycle-to-cycle variability to be small. Similarly having a pristine single crystal monolayer could mitigate any effects due to grain boundaries or other pinning centers that might cause variability in threshold voltage for the fire function.

Vertical MoS_2 devices using monolayer graphene as bottom contact and Ni top electrode have also been used to demonstrate integrate-fire function [34]. In this case, the MoS_2 layer thickness was of the order of 20 nm and polycrystalline. Combining such a vertical device with a capacitor enables self-sustaining oscillations in current output across the threshold switching device. The authors of this study found that the neuronal behavior was only observed when there was ambient oxygen and not in vacuum. This suggests contamination of the device with oxygen species that serve as mobile ions that may form and disrupt filaments resulting in sharp conductance jumps. Unlike the previous example, here the filament is not arising from the electrode material but rather an environmental gas species that can get adsorbed into the switching medium.

While the examples of neurons with 2D materials are still limited to date, the mechanisms supporting integrate-fire function can in principle be extended to various systems in this family. Mobile ion species that can migrate in the lattice, easily forming current shunts, offer one route that has been tested extensively in various oxide and other bulk systems. In layered semiconductors if the ion mobility could be enhanced in anisotropic manner, they can be exploited for fast filament formation across a channel to reach threshold at smaller voltages. If the filament can dissociate on its own when the stimulus is removed, the device can reset without the need for additional electrical energy input. Microstructure engineering using grain boundaries could enhance the migration kinetics at the same time creating complexity in hysteresis and stochasticity. On the other hand, if probabilistic switching is desirable (as will be discussed in Chapter 7), such complex features enable each individual device to be a rich source of physical phenomena. We note to the reader that judicious choice of switching mechanisms (i.e., extrinsic diffusive impurities, electrode dissolution, intrinsic structural distortions, and Joule heating) are therefore important beyond just the material system.

4.7 Spintronic Neurons

Figure 4.24 depicts different spintronic device structures that can offer a direct mapping to neuron functionalities with varying degrees of bio-fidelity. We first begin with a standard mono-domain MTJ where the device is initialized to the anti-parallel (AP) state and driven by a constant bias current equal to the threshold current required to switch the device to the parallel (P) state. Therefore, a synaptic current provided in addition to the bias current can either drive the device reliably to the P state or ensure it does not switch from the AP state, thereby emulating a "step" transfer function [35]. In addition to accuracy limitations associated with "step" transfer function neurons, the high bias current requirement limits the energy efficiency of the overall system. It is worth pointing out here that this same device can mimic enhanced functionalities like probabilistic neurons, when the internal time domain dynamics in the presence of thermal noise is considered. This will be discussed later.

One mechanism to reduce the high bias current requirement is to initialize the device in an unstable state. This can be achieved in a spin–orbit torque-based device where an MTJ lies on top of a heavy-metal (HM) layer [36]. Given the magnet has perpendicular magnetic anisotropy, a "Preset" current flowing through the HM layer will inject in-plane spins to the magnet lying on top, thereby orienting the device along the "hard-axis." If the "Preset" pulse is followed by the resultant synaptic current flowing through the MTJ before the magnet relaxes back to either of the two stable states, the direction of the current will dictate the final state of the MTJ, thereby realizing a "step" neuron functionality. Other implementations utilizing lateral spin valve-based structures (shown in the figure) and oscillators have also been explored in literature [37, 38].

The non-spiking neuron functionality is based on the domain wall motion-based MTJ structure discussed in Chapter 3. The same device can be interfaced with a reference MTJ forming a resistive divider [39]. The neuron operation takes place

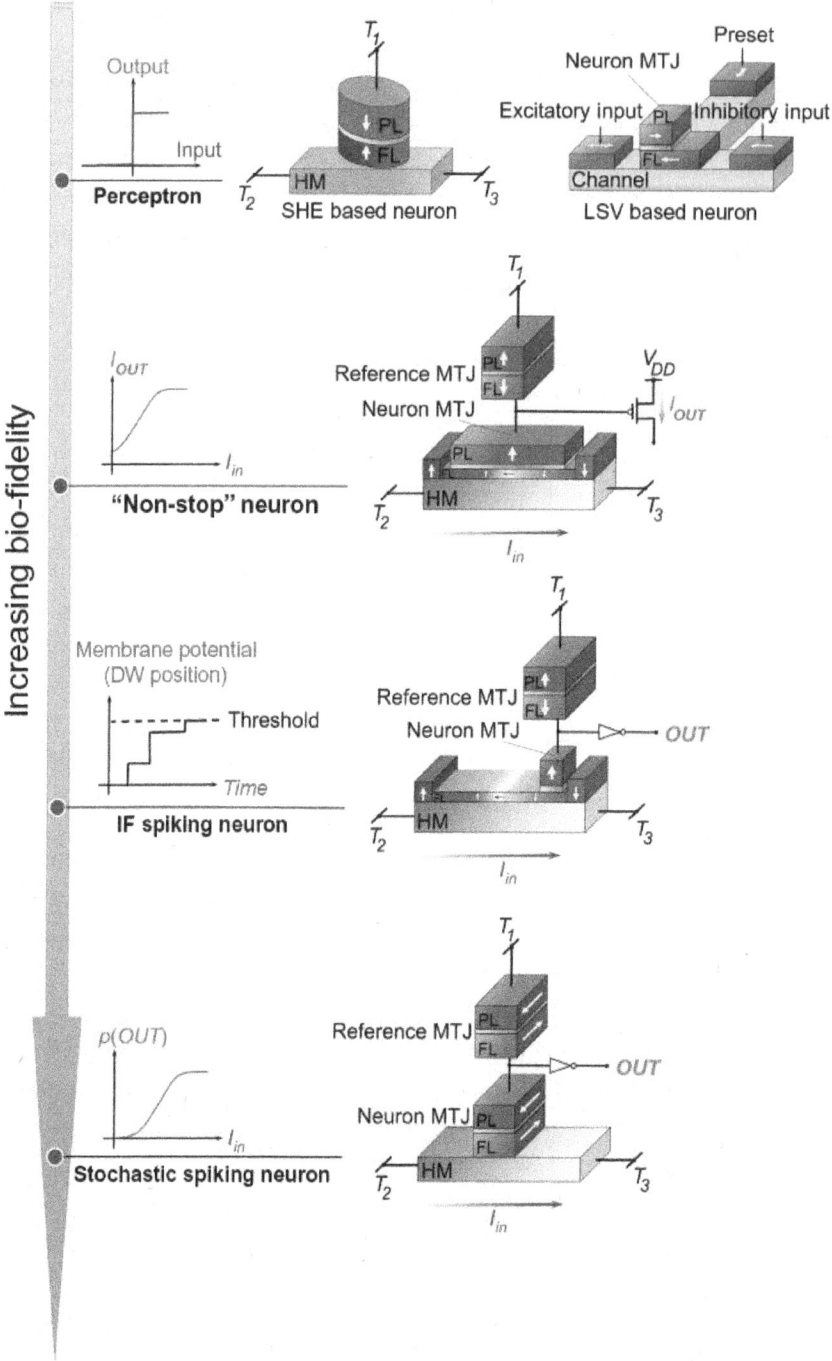

Figure 4.24 Various spintronic neuron device structures with increasing degrees of bio-fidelity – perceptron, "non-step" neuron, integrate-fire spiking neuron, and stochastic spiking neuron [49]. Adapted with permission from [49]. 2017 © AIP Publishing.

in three steps. Initially, the neuron state is written by passing a current through the underlying HM layer, which programs the domain wall position. It is worth pointing out here that spin neurons are magneto-metallic devices requiring very low currents for switching. Also, due to the decoupled write–read current paths associated with the device structure, the synaptic current flows through a low-resistance HM layer. These device design considerations will play an important role in low-cost peripheral overhead for neuromorphic system design involving the interfacing of synaptic crossbar arrays with spin-based neurons. After the neuron state has been written, the device state is read by using the resistive divider structure. The oxide thickness of the two MTJs can be chosen such that the read current does not disturb the state of the neuron MTJ. The resistive divider drives a transistor in the saturation regime (to ensure that the output current of the transistor does not depend on the interfaced fan-out synapses). With increase in the magnitude of the input write current through the HM layer, the DW displaces proportionally to the right and causes reduction in the MTJ resistance due to decrease in the proportion of AP domain in the free layer of the device. This causes a proportionate linear increase in the output current provided by the transistor. After the read operation, the domain wall position is reinitialized to the left edge of the free layer by passing a current through the HM layer in the opposite direction.

A very similar device-circuit structure can be used to mimic the functionality of an Integrate-Fire spiking neuron [40]. Considering the DW position to be analogous to the membrane potential of a spiking neuron, incoming current pulses through the HM can be deemed to be input spikes which causes a proportionate shift in the membrane potential or the DW location. If we consider the read MTJ to be on the extreme right edge of the free layer of the magnet, instead of the entire free layer, then the neuron spiking behavior can be thought of as the timestep when the DW reaches the extreme edge of the free layer and triggers the output inverter interfaced with the resistive divider. A constant leak functionality can be achieved by passing a current through the HM in the opposite direction at each timestep.

While the DW neuron provides the flexibility of implementing various neuron functionalities, it is limited by fabrication challenges as well as non-idealities like domain wall pinning at defect sites. Let us refocus our attention back to the mono-domain MTJ and approach the neuron functionality from a complementary perspective. The inherent magnetization dynamics of the MTJ has certain resemblances with the temporal dynamics of a leaky-integrate-fire spiking neuron [41]. Such dynamics when coupled with thermal noise at nonzero temperatures can be formulated as a probabilistic neuron where the probability of switching the MTJ varies in a nonlinear sigmoid fashion with the magnitude of the input current [41, 42]. Operation of the neuron occurs in a similar fashion where write, read, and reset phases are used per timestep of operation. Recent works [43, 44] have also explored such devices in the superparamagnetic regime where the device switches spontaneously between the P and AP states and the time-averaged magnetization varies in a similar nonlinear fashion with the input current magnitude. However, such devices require careful design space optimization with respect to peripheral design as they are highly sensitive to non-idealities [43]. Superparamagnetic devices are attractive for temporal encoding in spiking neural networks [45] as well as solving other unconventional computing paradigms like combinatorial optimization problems [46]. While we discussed a few examples of neuron device structures, other device structures utilizing different material stacks [47] as well as functionalities like oscillatory activities [48] are also being explored [49].

4.8 Photonic Neurons

In Chapter 3, we discussed a possible photonic synapse implementation [50–52] where a nonvolatile phase-change memory ($Ge_2Sb_2Te_5$ – GST) device was switched using sub-nanosecond optical pulses [52]. In this section, we will revisit the underlying principles of operation of the device and demonstrate how similar concepts can be utilized to construct a spiking neuron.

Figure 4.25 depicts a photonic neuron [52, 53] where the synaptic contributions are summed up in the "Integration Unit" followed by a thresholding operation in the "Firing Unit." Since trained synaptic weights in a network can be both positive and negative, positive and negative synaptic contributions are separated out through two individual add-drop ring resonators, each having its own GST element. The neuron operation occurs in a temporal fashion where each timestep can be divided into a "write" stage followed by a "read" stage. Upon application of "write" pulses, the GST elements start changing their state from the baseline crystalline state. After the "write" stage, the states of the GST elements are "read" using the transmission at the "through" and "drop" ports of each ring resonator. With partial amorphization of the GST material, transmission at the "through" ("drop") port decreases (increases). Therefore, the "drop" port is tapped out for positive synaptic contribution while the "through" port is utilized for the negative synaptic contribution. The two ports are connected to an interferometer, the output of which is analogous to the membrane potential of an integrate-fire spiking neuron. The interferometer output is then interfaced with an amplifier, a circulator, and a GST equipped waveguide to implement the thresholding operation. During the neuron's "read" state, the amplifier and circulator output begins to amorphize the GST element on the rectangular waveguide. Once the amplifier output is sufficient to completely amorphize the GST after application of several synaptic inputs, an output spike is generated. After the spiking operation, the neuron can be "reset" by resetting the states of all the devices to their initial states.

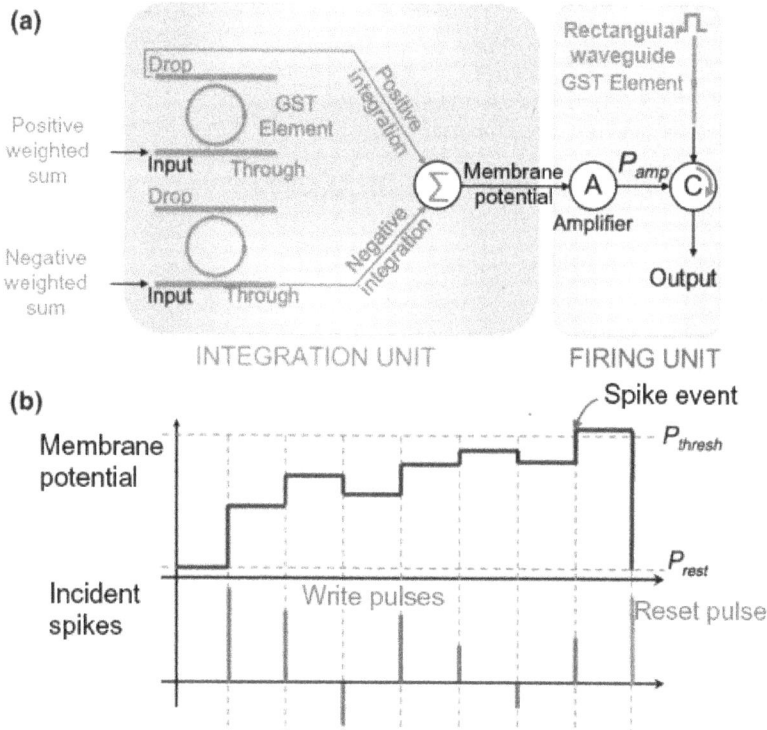

Figure 4.25 (a) Design of a photonic spiking neuron based on ring resonators with GST elements. The "Integration Unit" performs synaptic integration, while the "Firing Unit" performs the thresholding operation. (b) Temporal evolution of membrane potential dynamics of the neuron which is represented by the interferometer output [52]. Adapted with permission from [52]. © 2019 American Physical Society.

References

[1] Ghosh-Dastidar, S. and Adeli, H., 2009. Spiking neural networks. *International Journal of Neural Systems*, 19(04), pp. 295–308.

[2] Diehl, P. U. and Cook, M., 2015. Unsupervised learning of digit recognition using spike-timing-dependent plasticity. *Frontiers in Computational Neuroscience*, 9, p. 99.

[3] Chicca, E., Stefanini, F., Bartolozzi, C. and Indiveri, G., 2014. Neuromorphic electronic circuits for building autonomous cognitive systems. *Proceedings of the IEEE*, 102(9), pp. 1367–1388.

[4] Sengupta, A., Banerjee, A. and Roy, K., 2016. Hybrid spintronic-CMOS spiking neural network with on-chip learning: Devices, circuits, and systems. *Physical Review Applied*, 6(6), p. 064003.

[5] Dang, B., Liu, K., Zhu, J., Xu, L., Zhang, T., Cheng, C., Wang, H., Yang, Y., Hao, Y. and Huang, R., 2019. Stochastic neuron based on IGZO Schottky diodes for neuromorphic computing. *APL Materials*, 7(7), pp. 071114-1–071114-5.

[6] Woo, J., Lee, D., Koo, Y. and Hwang, H., 2017. Dual functionality of threshold and multilevel resistive switching characteristics in nanoscale HfO_2-based RRAM devices for artificial neuron and synapse elements. *Microelectronic Engineering*, 182, pp. 42–45.

[7] Mehonic, A. and Kenyon, A. J., 2016. Emulating the electrical activity of the neuron using a silicon oxide RRAM cell. *Frontiers in Neuroscience*, 10, p. 57.

[8] Funck, C., Menzel, S., Aslam, N., Zhang, H., Hardtgen, A., Waser, R. and Hoffmann-Eifert, S., 2016. Multidimensional simulation of threshold switching in NbO_2 based on an electric field triggered thermal runaway model. *Advanced Electronic Materials*, 2(7), p. 1600169.

[9] Goodwill, J. M., Sharma, A. A., Li, D., Bain, J. A. and Skowronski, M., 2017. Electro-thermal model of threshold switching in TaO x-based devices. *ACS Applied Materials & Interfaces*, 9(13), pp. 11704–11710.

[10] Mulaosmanovic, H., Chicca, E., Bertele, M., Mikolajick, T. and Slesazeck, S., 2018. Mimicking biological neurons with a nanoscale ferroelectric transistor. *Nanoscale*, 10(46), pp. 21755–21763.

[11] Arimoto, Y. and Ishiwara, H., 2004. Current status of ferroelectric random-access memory. *Mrs Bulletin*, 29(11), pp. 823–828.

[12] Chen, C., Yang, M., Liu, S., Liu, T., Zhu, K., Zhao, Y., Wang, H., Huang, Q. and Huang, R., 2019, June. Bio-inspired neurons based on novel leaky-FeFET with ultra-low hardware cost and advanced functionality for all-ferroelectric neural network. In *2019 Symposium on VLSI Technology* (pp. T136–T137). IEEE.

[13] Gibertini, P., Fehlings, L., Lancaster, S., Duong, Q.T., Mikolajick, T., Dubourdieu, C., Slesazeck, S., Covi, E. and Deshpande, V., 2022, October. A ferroelectric tunnel junction-based integrate-and-fire neuron. In *2022 29th IEEE International Conference on Electronics, Circuits and Systems (ICECS)* (pp. 1–4). IEEE.

[14] Garcia, V. and Bibes, M., 2014. Ferroelectric tunnel junctions for information storage and processing. *Nature Communications*, 5(1), p. 4289.

[15] Dagotto, E. and Tokura, Y., 2008. Strongly correlated electronic materials: Present and future. *Mrs Bulletin*, 33(11), pp. 1037–1045.

[16] Zhou, Y. and Ramanathan, S., 2013. Correlated electron materials and field effect transistors for logic: A review. *Critical Reviews in Solid State and Materials Sciences*, 38(4), pp. 286–317.

[17] Ruzmetov, D., Heiman, D., Claflin, B. B., Narayanamurti, V. and Ramanathan, S., 2009. Hall carrier density and magnetoresistance measurements in thin-film vanadium dioxide across the metal-insulator transition. *Physical Review B – Condensed Matter and Materials Physics*, 79(15), p. 153107.

[18] Lin, J., Guha, S. and Ramanathan, S., 2018. Vanadium dioxide circuits emulate neurological disorders. *Frontiers in Neuroscience*, 12, p. 856.

[19] Yi, W., Tsang, K. K., Lam, S. K., Bai, X., Crowell, J. A. and Flores, E. A., 2018. Biological plausibility and stochasticity in scalable VO_2 active memristor neurons. *Nature Communications*, 9(1), p. 4661.

[20] Gao, L., Chen, P. Y. and Yu, S., 2017. NbOx based oscillation neuron for neuromorphic computing. *Applied Physics Letters*, 111(10), pp. 103503-1–103503-4.

[21] Liu, X., Li, S., Nandi, S. K., Venkatachalam, D. K. and Elliman, R. G., 2016. Threshold switching and electrical self-oscillation in niobium oxide films. *Journal of Applied Physics*, 120(12), pp. 124102-1–124102-10.

[22] Shi, Y., Duwel, A. E., Callahan, D. M., Sun, Y., Hong, F. A., Padmanabhan, H., Gopalan, V., Engel-Herbert, R., Ramanathan, S. and Chen, L. Q., 2021. Dynamics of voltage-driven oscillating insulator-metal transitions. *Physical Review B*, 104(6), p. 064308.

[23] Lin, J., Alam, K., Ocola, L., Zhang, Z., Datta, S., Ramanathan, S. and Guha, S., 2017, December. Physics and technology of electronic insulator-to-metal transition (E-IMT) for record high on/off ratio and low voltage in device applications. In *2017 IEEE International Electron Devices Meeting (IEDM)* (pp. 23–24). IEEE.

[24] Amer, S., Hasan, M. S., Adnan, M. M. and Rose, G. S., 2019, August. Design considerations for insulator metal transition based artificial neurons. In *2019 IEEE 62nd International Midwest Symposium on Circuits and Systems (MWSCAS)* (pp. 1131–1134). IEEE.
[25] Yu, H., Islam, A. N., Mondal, S., Sengupta, A. and Ramanathan, S., 2022. Switching dynamics in vanadium dioxide-based stochastic thermal neurons. *IEEE Transactions on Electron Devices*, 69(6), pp. 3135–3141.
[26] Park, T. J., Selcuk, K., Zhang, H. T., Manna, S., Batra, R., Wang, Q., Yu, H., Aadit, N. A., Sankaranarayanan, S. K., Zhou, H. and Camsari, K. Y., 2022. Efficient probabilistic computing with stochastic perovskite nickelates. *Nano Letters*, 22(21), pp. 8654–8661.
[27] Stoliar, P., Tranchant, J., Corraze, B., Janod, E., Besland, M. P., Tesler, F., Rozenberg, M. and Cario, L., 2017. A leaky-integrate-and-fire neuron analog realized with a Mott insulator. *Advanced Functional Materials*, 27(11), p. 1604740.
[28] Kalcheim, Y., Camjayi, A., Del Valle, J., Salev, P., Rozenberg, M. and Schuller, I. K., 2020. Non-thermal resistive switching in Mott insulator nanowires. *Nature Communications*, 11(1), p. 2985.
[29] Harikesh, P. C., Yang, C. Y., Tu, D., Gerasimov, J. Y., Dar, A. M., Armada-Moreira, A., Massetti, M., Kroon, R., Bliman, D., Olsson, R. and Stavrinidou, E., 2022. Organic electrochemical neurons and synapses with ion mediated spiking. *Nature Communications*, 13(1), p. 901.
[30] Sarkar, T., Lieberth, K., Pavlou, A., Frank, T., Mailaender, V., McCulloch, I., Blom, P. W., Torricelli, F. and Gkoupidenis, P., 2022. An organic artificial spiking neuron for in situ neuromorphic sensing and biointerfacing. *Nature Electronics*, 5(11), pp. 774–783.
[31] Tischler, V., Dudek, P., Wijekoon, J., Majewski, L. A., Takeda, Y., Tokito, S. and Turner, M. L., 2023. An integrate-and-fire neuron circuit made from printed organic field-effect transistors. *Organic Electronics*, 113, p. 106685.
[32] Hosseini, M. J. M., Donati, E., Yokota, T., Lee, S., Indiveri, G., Someya, T. and Nawrocki, R.A., 2020. Organic electronics Axon-Hillock neuromorphic circuit: Towards biologically compatible, and physically flexible, integrate-and-fire spiking neural networks. *Journal of Physics D: Applied Physics*, 54(10), p. 104004.
[33] Hao, S., Ji, X., Zhong, S., Pang, K. Y., Lim, K. G., Chong, T. C. and Zhao, R., 2020. A monolayer leaky integrate-and-fire neuron for 2D memristive neuromorphic networks. *Advanced Electronic Materials*, 6(4), p. 1901335.
[34] Kalita, H., Krishnaprasad, A., Choudhary, N., Das, S., Dev, D., Ding, Y., Tetard, L., Chung, H. S., Jung, Y. and Roy, T., 2019. Artificial neuron using vertical MoS2/graphene threshold switching memristors. *Scientific Reports*, 9(1), p. 53.
[35] Sengupta, A. and Roy, K., 2015, July. Spin-transfer torque magnetic neuron for low power neuromorphic computing. In *2015 International Joint Conference on Neural Networks (IJCNN)* (pp. 1–7). IEEE.
[36] Sengupta, A., Choday, S. H., Kim, Y. and Roy, K., 2015. Spin orbit torque based electronic neuron. *Applied Physics Letters*, 106(14), pp. 143701-1–143701-5.
[37] Sharad, M., Panagopoulos, G. and Roy, K., 2012, June. Spin neuron for ultra low power computational hardware. In *70th device research conference* (pp. 221–222). IEEE.
[38] Arai, H. and Imamura, H., 2018. Neural-network computation using spin-wave-coupled spin-torque oscillators. *Physical Review Applied*, 10(2), p. 024040.
[39] Sengupta, A., Shim, Y. and Roy, K., 2016. Proposal for an all-spin artificial neural network: Emulating neural and synaptic functionalities through domain wall motion in ferromagnets. *IEEE Transactions on Biomedical Circuits and Systems*, 10(6), pp. 1152–1160.

[40] Sengupta, A. and Roy, K., 2016. A vision for all-spin neural networks: A device to system perspective. *IEEE Transactions on Circuits and Systems I: Regular Papers*, *63*(12), pp. 2267–2277.
[41] Sengupta, A., Panda, P., Wijesinghe, P., Kim, Y. and Roy, K., 2016. Magnetic tunnel junction mimics stochastic cortical spiking neurons. *Scientific Reports*, *6*(1), p. 30039.
[42] Sengupta, A., Parsa, M., Han, B. and Roy, K., 2016. Probabilistic deep spiking neural systems enabled by magnetic tunnel junction. *IEEE Transactions on Electron Devices*, *63*(7), pp. 2963–2970.
[43] Liyanagedera, C. M., Sengupta, A., Jaiswal, A. and Roy, K., 2017. Stochastic spiking neural networks enabled by magnetic tunnel junctions: From nontelegraphic to telegraphic switching regimes. *Physical Review Applied*, *8*(6), p. 064017.
[44] Pervaiz, A. Z., Datta, S. and Camsari, K. Y., 2019, November. Probabilistic computing with binary stochastic neurons. In *2019 IEEE BiCMOS and Compound Semiconductor Integrated Circuits and Technology Symposium (BCICTS)* (pp. 1–6). IEEE.
[45] Yang, K. and Sengupta, A., 2020. Stochastic magnetoelectric neuron for temporal information encoding. *Applied Physics Letters*, *116*(4), pp. 043701-1–043701-5.
[46] Aadit, N. A., Grimaldi, A., Carpentieri, M., Theogarajan, L., Finocchio, G. and Camsari, K. Y., 2021, December. Computing with invertible logic: Combinatorial optimization with probabilistic bits. In *2021 IEEE International Electron Devices Meeting (IEDM)* (pp. 40–43). IEEE.
[47] Kurenkov, A., DuttaGupta, S., Zhang, C., Fukami, S., Horio, Y. and Ohno, H., 2019. Artificial neuron and synapse realized in an antiferromagnet/ferromagnet heterostructure using dynamics of spin–orbit torque switching. *Advanced Materials*, *31*(23), p. 1900636.
[48] Khymyn, R., Lisenkov, I., Voorheis, J., Sulymenko, O., Prokopenko, O., Tiberkevich, V., Akerman, J. and Slavin, A., 2018. Ultra-fast artificial neuron: generation of picosecond-duration spikes in a current-driven antiferromagnetic auto-oscillator. *Scientific Reports*, *8*(1), p. 15727.
[49] Sengupta, A. and Roy, K., 2017. Encoding neural and synaptic functionalities in electron spin: A pathway to efficient neuromorphic computing. *Applied Physics Reviews*, *4*(4), pp. 041105-1–041105-23.
[50] Ríos, C., Youngblood, N., Cheng, Z., Le Gallo, M., Pernice, W. H., Wright, C. D., Sebastian, A. and Bhaskaran, H., 2019. In-memory computing on a photonic platform. *Science Advances*, *5*(2), p. eaau5759.
[51] Stegmaier, M., Ríos, C., Bhaskaran, H., Wright, C. D. and Pernice, W. H., 2017. Nonvolatile all-optical 1 × 2 switch for chipscale photonic networks. *Advanced Optical Materials*, *5*(1), p. 1600346.
[52] Chakraborty, I., Saha, G. and Roy, K., 2019. Photonic in-memory computing primitive for spiking neural networks using phase-change materials. *Physical Review Applied*, *11*(1), p. 014063.
[53] Chakraborty, I., Saha, G., Sengupta, A. and Roy, K., 2018. Toward fast neural computing using all-photonic phase change spiking neurons. *Scientific Reports*, *8*(1), p. 12980.

5

Examples of Applications in Artificial Neural Networks

5.1 Silicon CMOS-Based Neural Networks

As discussed earlier, the core computationally expensive operation required to enable hardware acceleration of neural networks is the dot-product operation of synaptic weights (stored in memory) with neuronal activations. Traditional von Neumann-based architectures suffer from memory transaction bottleneck, thereby resulting in reduced throughput and enhanced energy consumption. "Near-memory" computing approaches such as those exploited in neuromorphic processors such as Intel Loihi [1] and IBM TrueNorth [2], divides the entire computing workload into smaller portions and physically colocates multiple memory and processing units together, thereby providing better localization of memory transactions.

Achieving "In-Memory" computing in digital CMOS SRAM memories requires certain modifications to the 6T-SRAM structure since the read and write ports are coupled. Two extra read transistors can be added to the 6T structure (as shown in Figure 5.1) to decouple the read and write bit-lines and word-lines. Boolean bit-wise NOR, NAND, and XOR operation between bits stored in the SRAM cells can be performed by simultaneous activation of the read word-lines (RWL) associated with the cells and readout of the read bit-line (RBL) using appropriate sensing circuitry [3, 4]. Addition and multiplication operations can be also performed by using peripheral low-cost adders [5]. Such designs are ideally suited for binary neural network hardware accelerators since dot products can be simplified to XOR operations with binary inputs and binary weights [6]. Analog implementations of "In-Memory" computing in SRAM cells have been also explored to improve the throughput of the dot-product computing process [7, 8].

Next, we shift gears from circuit aspects to architectural implementations. Neuromorphic computing architectures leverage the spatial and temporal sparsity of spike-based computing and communication in SNNs [9]. For instance, Loihi

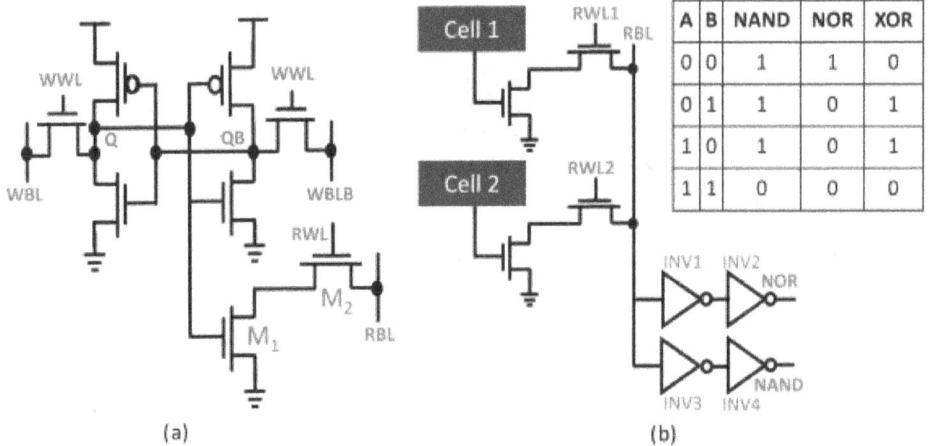

Figure 5.1 (a) 8T SRAM cell for implementing "In-Memory" computing. (b) Single-ended sensing of NAND/NOR operations using gated skewed inverters [4]. Adapted with permission from [4]. Copyright © 2018, IEEE.

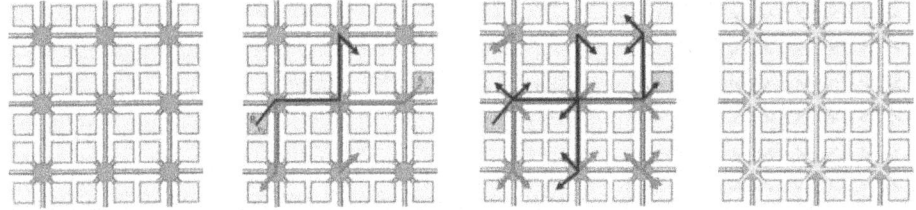

Figure 5.2 Mesh architecture used in Loihi chip [11]. Adapted with permission from [11]. Copyright © 2018, IEEE.

utilizes a hierarchical mesh architecture and asynchronous data communication [10, 11] (see Figure 5.2) to leverage sparsity effects. The chip has multiple cores that operate independently, and each has its own sense of spiking (clocking) frequency and timing.

The data communication between cores is asynchronous. Upon a spiking event, the network-on-chip (NoC) broadcasts a spike message to only the fan-out cores. Upon processing of a spike by the slowest core, a synchronization event between neighboring cores takes place to ensure reliable delivery and receipt of spikes. The local dataflow control in the asynchronous Loihi architecture, where each core operates at different frequencies, facilities application-dependent varying timescale support needed for SNNs along with facilitating back-end timing closure. Other CMOS-based neuromorphic architectures like TrueNorth consider the computing cores to be synchronous with asynchronous data communication between the cores [9]. All the cores therefore operate in parallel with asynchronous event-driven activation of fan-out cores in case of a spiking event.

Asynchronous data communication in large-scale neuromorphic chips require high-speed, time-multiplexed asynchronous circuits that consist of a transceiver to parallelly read and write spikes and a router to transmit spikes. A commonly used asynchronous communication protocol is termed address event representation (AER) [12]. The signals used in AER protocol convey information of inter-spike intervals in digital format along with the transmitting node addresses. A look-up table-based approach with source and destination pairs of addresses can be maintained to ensure proper routing of spikes. However, naïve implementation of such a look-up table can involve high memory overhead. Specialized connectivity patterns of neurons in a network can be considered while developing routing tables (like multistage [13] and hierarchical [14] schemes) to reduce memory requirements for implementation of reconfigurable SNNs. Many of the above architectural considerations are equally valid and have been also adopted for post-CMOS hardware accelerators. CMOS-based neuromorphic chips like Loihi [15] have shown significant hardware efficiency in various applications ranging from nearest neighbor search, graph search, stochastic constrained optimization problems to real-world robotic, and simultaneous localization and mapping (SLAM) applications. These preliminary demonstrations suggest significant promise for implementing brain-inspired intelligence in hardware in improving energy efficiency.

5.2 Correlated Electron Semiconductor-Based Neural Networks

The electrically driven insulator–metal transition (IMT) in correlated electron semiconductors is a foundational property utilized to construct artificial neural circuits. While in Chapters 3 and 4 we have discussed the working mechanisms and properties of individual devices that can act as neurons or synapses, here we focus on interconnected devices that form elementary circuit-level building blocks. Proof-of-concept neuromorphic networks have been demonstrated with various correlated electron semiconductors, and their performance has been evaluated via combination of experiment and modeling for tasks such as pattern recognition. One such study used a combination of pristine VO_2 and H-doped VO_2: Neuronal and synaptic devices were connected to design a feedforward circuit [16]. Pristine VO_2 acts as a threshold switch that can serve as artificial neuron via periodic voltage oscillations. When it is doped with hydrogen, the electron donated from hydrogen destabilizes the insulating state thereby reducing the insulating state resistance. Further, the charged proton can migrate in the lattice under electric fields enabling multitude of nonvolatile resistance states.

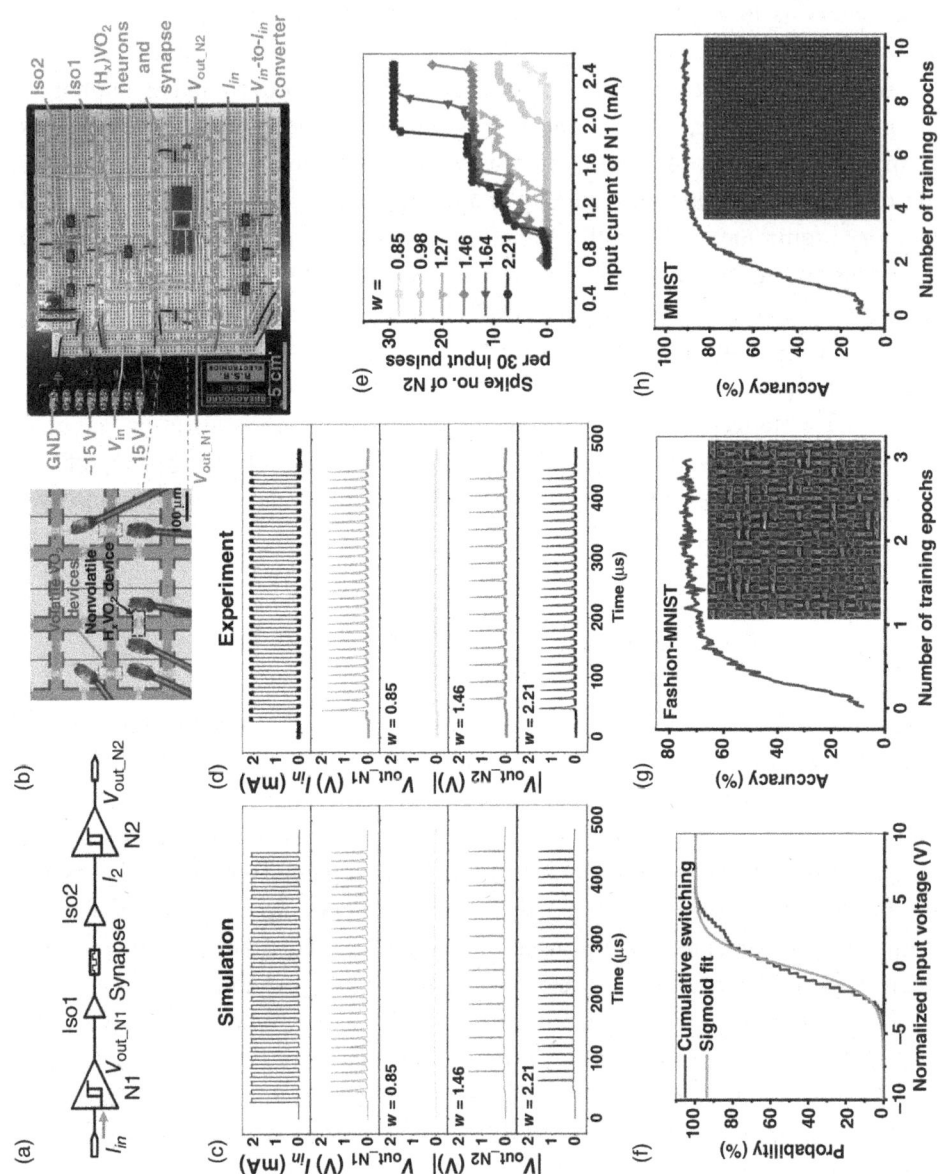

Using locally defined catalytic electrodes for hydrogen doping via spillover, it then becomes possible to create spatial configurations of neurons and synapses within the same parent semiconductor layer on a single chip. The neuron properties are governed by the two-terminal device geometry while the synapse weight can be programmed by voltage and pulse width. Figure 5.3(a) shows schematic of such a connected circuit incorporating vanadium dioxide-based switches. The isolators are included to maintain proper voltage balance in the circuit. Figure 5.3 (b) shows optical image of the circuit hardware. Figure 5.3(c) and (d) shows simulated and experimental output voltage waveforms when the synaptic weight is varied. The periodicity of spiking is clearly dependent on the synapse weight. Figure 5.3(e) shows how the probability of spiking behavior of neuron 2 is dependent on both the input current to neuron 1 and the synaptic weight. Neuron 2 can therefore behave as an excitatory or inhibitory neuron. Essentially, the spiking behavior can be significantly influenced by the input current like what happens in a biological circuit. Figure 5.3(f) shows the cumulative dynamics of phase switching fitted to a sigmoidal profile. Figure 5.3(g) and (h) shows simulated performance of the network for pattern recognition for different databases. Such motifs can be expanded into larger scale networks on chip as the fabrication process is compatible with traditional lithographic fabrication of electronic devices. The properties of the neuron and synaptic devices can be fine-tuned depending on the circuit. For instance, the composition of the oxide locally can be modified by synthesis or post-annealing conditions to vary the oxygen sub-lattice stoichiometry which can in turn control the threshold voltage for spiking. Similarly, hydrogen doping of the layer can be controlled by time-temperature annealing envelopes to incorporate different amounts of the dopants in the lattice. Post-doping, the distribution of the dopants and the resulting electrical resistance can be modulated by electric field profile. Hence, several process and design variables can be utilized for optimizing the neuromorphic circuits.

To further illustrate how the stoichiometry control of the IMT layer can influence circuit dynamics, we can consider two VO_2 neurons connected through a

Figure 5.3 (a) Drawing of a feedforward neural circuit utilizing a prototypical correlated oxide, VO_2. (b) Photo of the hardware chip and breadboard. (c and d) Shows simulated and measured output waveforms of neuron N2 as the synaptic weight is varied. W represents the synaptic weight and is defined as the absolute value of the gain of the inverting amplifier. (e) Probability of spiking of neuron N2 based on input current of neuron N1 gated by the synapse. The probability is the ratio of the spike number divided by 30 which represents the number of input current pulses. (f) Switching dynamics of the stochastic neurons and a sigmoidal fit that can be implemented into network-level simulations. (g and h) Shows the performance metrics of the simulated network utilizing the measured device characteristics for two different databases [16]. Adapted with permission from [16] under the Creative Commons Attribution 4.0 license.

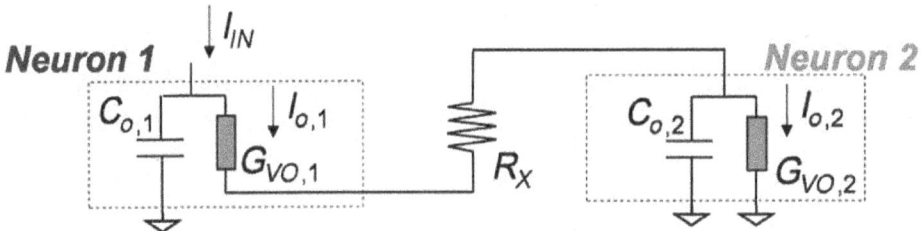

Figure 5.4 Schematic of a monosynaptic circuit comprising two IMT threshold switching neurons connected by a synaptic weight [17]. Adapted with permission from [17] under the Creative Commons Attribution 4.0 license.

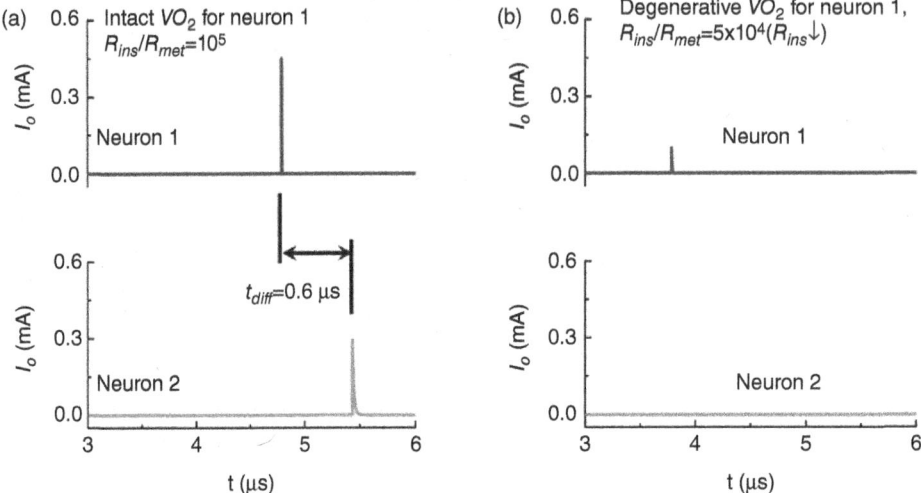

Figure 5.5 (a) Neuron 1 firing results in generation of action potential in neuron 2 after a time delay corresponding to signal transfer timescale. (b) Neuron 1 firing does not result in firing of neuron 2 since the ground-state resistance of neuron 1 is decreased compared to (a). The current leaks before it can reach neuron 2 resulting in failure to reach threshold needed for firing [17]. Adapted with permission from [17] under the Creative Commons Attribution 4.0 license.

variable resistor that acts as a synaptic weight as shown in Figure 5.4 [17]. If the two VO_2 devices are fabricated with near-ideal stoichiometry resulting in similar high-resistance state (HRS, corresponding to the insulating phase), when neuron 1 fires, the signal is transferred to the subsequent neuron after a time delay causing the neuron 2 to fire as shown in Figure 5.5. If, however, neuron 1 has a lower ground-state resistance due to non-stoichiometry in the oxygen sub-lattice, then the signal is not transferred with sufficient current amplitude such that neuron 2 does not fire. Control over the composition is a powerful method to tune the signal propagation characteristics in networks and reminiscent of the role of myelin sheaths in ensuring signal transmission in biological neurons. Recall that myelin sheaths provide electrical insulation for biological neurons resulting in high-fidelity signal transmission. At the same time, it points to the crucial role of controlled synthesis protocols needed to maintain robust information transfer using phase transition materials. This will become more apparent in the discussion that follows on oscillatory networks.

The hysteretic current–voltage characteristics seen in IMT-based devices also enables oscillatory neural networks (ONNs). These networks are inspired by observations in neuroscience concerning brain waves and synchronization behavior in neural circuits [18]. In this approach, the relative phase of the oscillators can be used for computing. Multiple VO_2 oscillators have been coupled with resistors (that act as synapses). By varying the time between input voltage, it is possible to induce delay between the connected devices as they begin the respective oscillations between insulating and metallic states [19]. The values of the synaptic resistors are established during the training stage using user-defined plasticity rules. To recognize patterns such as letters or digits, the oscillator devices can be referenced to a control device and pixel values such as white or black can be determined by comparison of the phase delay (0° or 180°), with respect to the reference oscillator. By varying the phase delay between these extreme values, it is possible to generate a gray scale of pixels where each pixel represents an oscillator. Up to 60 oscillators have been demonstrated via circuit simulations to recognize patterns and shown to be robust even for noisy data [19]. The energy for operations scale with device geometry as the threshold voltage for activation reduces with gap between electrodes. Similarly, it can be argued that the frequency dependence can be improved by choice of the neuronal device and the capacitors that are connected to realize the oscillations. While projected performance of such ONN circuits for image analysis has been predicted to outperform off-the-shelf GPUs, practical limitations include control of device-to-device (D2D) variability that severely constrains the scalability of such emerging technologies [20].

As described in Chapters 3 and 4, there exist several materials that display IMT characteristics. Although not all are equal in terms of the magnitude of the conductivity jump across the transition and ability to operate near room temperature. NbO_2 is another system that has garnered much interest in this context. Like VO_2, the binary oxide can be grown nearly stoichiometric or as NbO_x with x deviating significantly from the nominal value of 2. The IMT behavior in this material has been utilized to demonstrate coupled switching devices that can act as neuristors to emulate Hodgkin–Huxley dynamics [21]. By careful selection of device and circuit parameters, it is possible to realize a range of spiking behaviors that have been noted in biological circuits. Further, by starting with a non-stoichiometric active layer, it is possible to realize multiple functions such as neuronal and synaptic characteristics within a single device. Starting with an amorphous NbO_x layer that is deposited by sputtering, it is possible to electroform the layer (electroforming

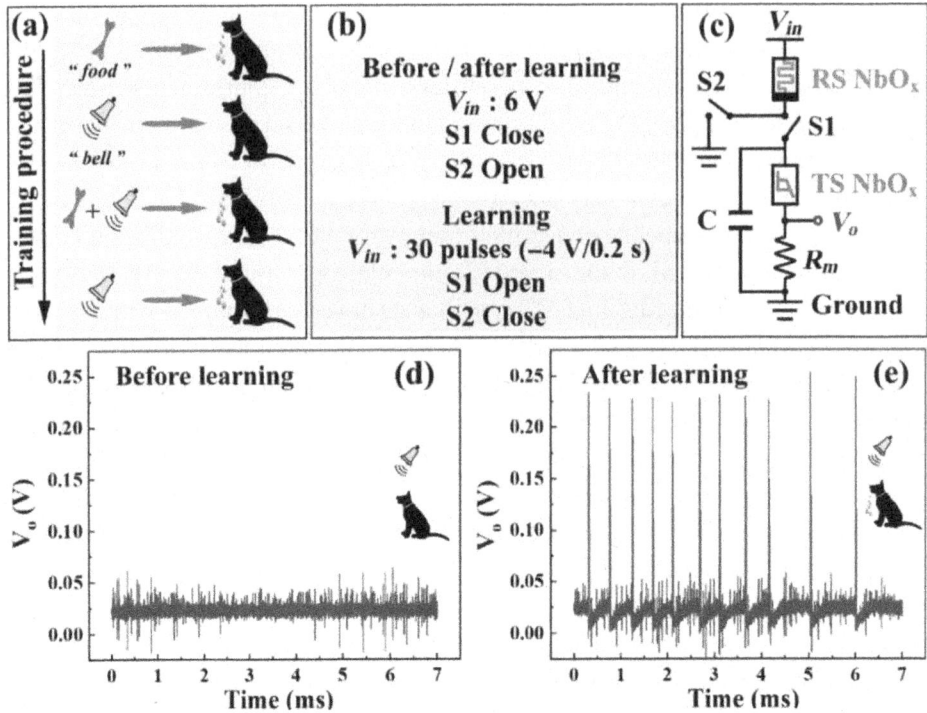

Figure 5.6 (a) Schematic of a Pavlovian dog associating between food and ringing of a bell; (b) electrical stimulus conditions for the learning process. (c) Circuit diagram showing connection between a threshold switch and synapse, both are fabricated with the same 50 nm NbO_x material, although within the channel region, the composition is different. (d and e) Depending on the resistance state of the synapse, the neuron device shows firing or not [22]. Adapted with permission from [22].

refers to applying different electrical bias to modify the channel electrical property) to act as a neuron or synapse.

Application of electrical stimuli results in migration of oxygen vacancies that can slowly relax back when the stimulus is removed. This enables them to operate as synapses with forgetting capability (i.e., transient memory). On the other hand, application of larger stimulus results in abrupt switching reminiscent of neuronal function [22]. The mechanisms for the two functions are rather distinct. In the former case, migration of oxygen vacancies results in variable resistance between the electrodes as discussed in Chapter 4. For threshold switching, the larger forming voltage results in formation of a local NbO_2 layer that helps realize a threshold switch. Thus, we have the two basic components to demonstrate a neural circuit using the Nb:O materials system. Figure 5.6 shows an example of learning made famous by Pavlov's experiments with dogs where in a stimulus originally not related to a response can be trained to be associative.

Figure 5.6(a) shows the schematic of the training process following the work of Pavlov, and Figure 5.6(b) describes the testing conditions. The circuit showing the connected devices is shown in Figure 5.6(c). When the synapse is at HRS, the neuron device does not oscillate (Figure 5.6(d)) while oscillations are noted (Figure 5.6(e)) when the synapse is in low-resistance state (LRS) after electrical inputs that represent training. Implementing the characteristics into larger scale simulations of spiking neural networks resulted in image recognition accuracy of over 85% after 100 training epochs and this increased to over 91% implementing the forgetting behavior of the resistive synapses as unimportant weights were gradually forgotten [22]. This concept of combining stoichiometric NbO_2 that acts as a threshold switch and non-stoichiometric oxide (NbO_x or other such as TaO_x) in series has been explored in a plethora of studies to demonstrate bioinspired neural circuits for image recognition, motion recognition and so forth.

The threshold switching NbO_x neurons can also be utilized for ONNs as discussed earlier in this chapter. By combining multiple neurons with memory resistors, it is possible to perform computing based on phase delay between the devices. The phase delay is induced by varying the time-delay between supplying the driving voltage to the various devices. Based on performance of individual devices, larger scale simulations have shown successful applications in edge detection [23]. As discussed in Section 5.1, it is important for the devices in the network to have rather similar oscillation characteristics such that any difference in phase arises entirely from the time-delay in signaling rather than extrinsic variability. We refer the reader to ref [24] for a more detailed discussion on approaches to implement learning rules for the coupling resistances in ONNs and constraints for scaling up ONNs.

So far, we have discussed networks where either the spiking behavior of the neurons coupled with resistors or capacitors can be utilized for computing via spike frequency (SNNs) or phase delay (ONNs) encoding. There are additional features in such threshold switches that could be considered for computing. For instance, when two NbO_x neurons are resistively coupled, it is possible to identify regimes of burst spiking. From a biological viewpoint, this roughly corresponds to mismatch in charging and discharging kinetics of sodium and potassium ion channels. The number of spikes in each burst period contains information and therefore can be utilized for pattern classification. Additional features such as interburst frequency may also be utilized for encoding information [25].

The examples described earlier utilized two-terminal devices to realize both neurons and synapses. It is also possible to utilize three-terminal synaptic transistors for constructing neural circuits. By modulating the chemical composition of the channel through addition or removal of point defects such as oxygen vacancies, the resistance states can be tuned in a nonvolatile manner. Perovskite nickelates such as $SmNiO_3$-based synaptic transistors have been wired with conventional electronics to emulate various forms of associative and nonassociative learning [26]. Electron-doped perovskite nickelates have also inspired novel network structures based on grow-when-required (GWR) architecture based on their reconfigurable properties in two-terminal architectures. Further, such devices that host combination of resistive and capacitive properties can be utilized for reservoir computing [27]. The class of correlated electron semiconductors are therefore a versatile family for design of artificial neural circuits that can function at room temperature and above with the capability of monolithic integration with silicon CMOS circuits. The spatial electronic phase inhomogeneity enables versatile neuromorphic function in both two and three-terminal device structures that is the subject of intense study currently.

5.3 Filamentary Switch-Based Neural Networks

Several experimental demonstrations and modeling of neural networks incorporating filamentary switching memristors have been reported in literature. We summarize a few examples and refer the reader to more comprehensive reviews for detailed aspects of each technology. One of the first examples are hybrid CMOS transistor-based neurons and filamentary synapses. As discussed in Chapter 3, silicon transistor-based synaptic devices require complex circuity and chip footprint whereas two-terminal devices with internal memory dynamics can serve as compact synapses. This has inspired several designs to integrate silicon neurons with filamentary switch-based synapses. Further, if the synapses can be fabricated as crossbar structures, then an array of such devices can be connected to neurons. Figure 5.7 shows an example of hybrid CMOS-memristor neural circuits [28].

The memristor array is comprised of a crossbar structure where metallic nanowires sandwich a memristor layer whose conductance can be modulated by voltage bias. The nanowire patterns are constructed such that the arrays connect with vias in the CMOS layer to distinct transistor gates. The memristor layer acts as a synapse with tunable conductance and hence determines how the current flows in the circuit. In practice, design and efficient operation of such hybrid circuits require either additional circuitry for matching the CMOS operational voltage/current conditions with that of the nonvolatile memory arrays and to avoid accidentally turning on/off adjacent devices due to the electrical stimulus that are shared across arrays. Nevertheless, such an approach demonstrates the feasibility to co-integrate emerging memory and hardware accelerator technologies with CMOS chips that are manufactured in semiconductor foundries [29]. Note that memristors need not always be nonvolatile and can be engineered to possess transient memory. Such devices can be readily co-integrated with nonvolatile synaptic memory technologies to demonstrate neuromorphic circuits.

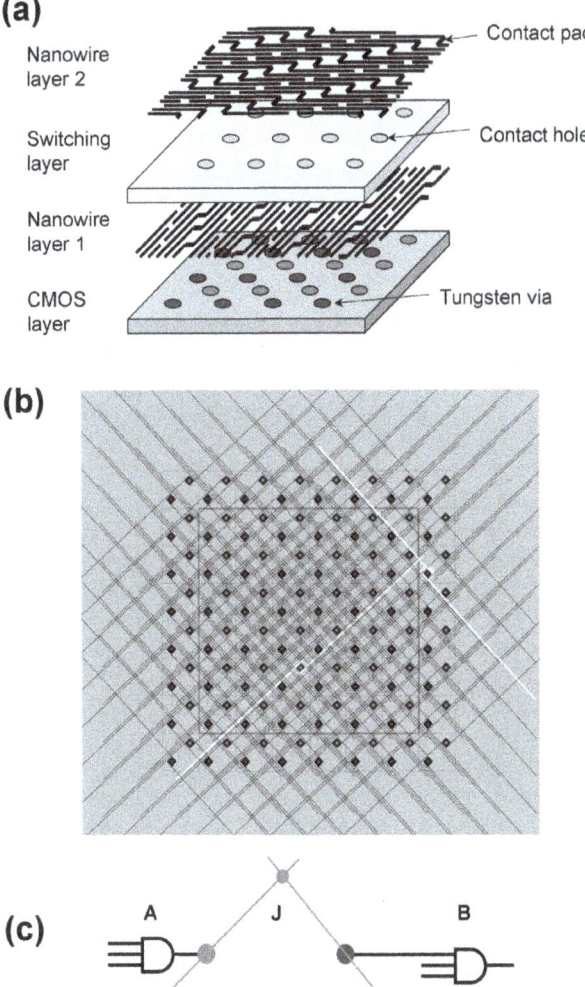

Figure 5.7 (a) Schematic showing CMOS chip integrated with memristor array. The vias are shown as black and dark gray dots. The nanowire layers 1 and 2, respectively, serve as electrodes that connect to transistor gates as shown in (c). (b) Shows the top view of nanowires forming crossbar structures. In (c), the transistor gates A and B are connected through memristor J. Depending on the conducting state of J, current flows through the wires [28]. Adapted with permission from [28]. Copyright © 2009 American Chemical Society.

5.4 Organic Electronic Neural Networks

Organic materials offer the advantages of biocompatibility, mechanical flexibility and large area scalable based fabrication processes. Several demonstrations of neuromorphic components such as artificial spiking neurons, synaptic devices have been reported to date. These devices can be connected to traditional

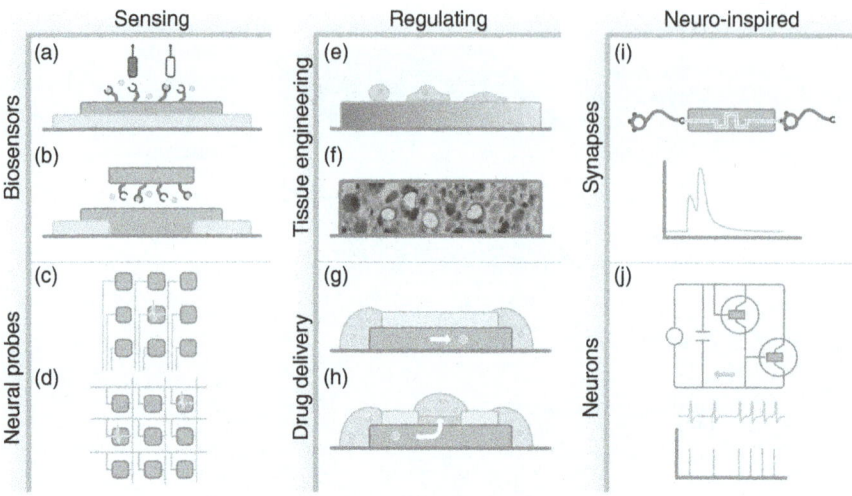

Figure 5.8 Organic semiconductors enable circuits for various bio-interfaced and bioinspired functions. (a–d) Devices and circuitry for sensing bio-molecules; increasing signal-noise ratio via functionalizing the gate electrode or channel and multiplexing. (e–h) Surfaces and 3D scaffolds for interfacing with cellular media, ion pumps and ion diodes for selective release of transmitter delivery. (i–j) Organic synapses and neuron circuitry for neuro-inspired computation. When interfaced with biological matter, incoming signals can be seamlessly processed with the artificial circuits enabling brain-machine interfaces [30]. Adapted with permission from [30].

transistors built with organic semiconductors for hybrid analog-digital circuits like their inorganic counterparts. Full neuromorphic circuits using organic semiconductors are still in their infancy at the time of writing this book. Note that due to their biocompatibility and low carrier mobility resulting in slow switching speeds, organic-material based circuits are also of interest in the context of bioelectronics and sensing in addition to computing. Figure 5.8 shows an overview of devices comprising organic materials for various functions involving interfacing with the environment and interpretation of signals [30].

Polyaniline (PANI) has been utilized to demonstrate perceptron circuits for pattern classification. PANI-based memristive devices with lithium perchlorate electrolyte mixed in water and polyethylene oxide were fabricated and three such devices were connected to an ammeter representing a neuron output. NAND and NOR logic functions were demonstrated with such a simple network. The potentiation–depression voltages between −0.2 V and 0.7 V are set by the redox characteristics of the PANI-electrolyte stack while the duration of the training pulses was determined by the relaxation of the conductance upon voltage stimulus and of the order of 100 s approximately [31].

Using polymers comprised of 2-(2-thienyl)-(3,4-ethylenedioxythiophene) TEDOT, organic electrochemical transistor (OECT) arrays have been fabricated

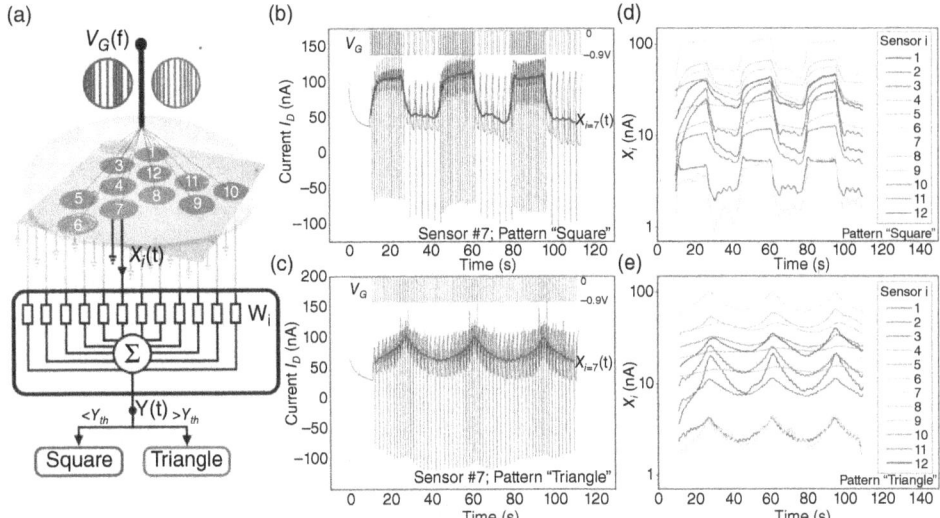

Figure 5.9 (a) Schematic of experiment combining 12 OECTs for the purpose of signal classification. (b and c) Current–time characteristics of an OECT to square and triangle voltage pulses of −900 mV amplitude and 1 s duration. (d and e) Response of 12 OECTs to the two voltage pulse profiles, respectively. There is an inherent frequency response of the OECT due to the nonequilibrium distribution of the ions in the semiconductor channel that screens the field. There is variability in both mean current and current modulation between devices that can be exploited for reservoir computing [32]. Adapted with permission from [32]. © 2018 WILEY-VCH Verlag GmbH & Co. KGaA, Weinheim.

toward the goal of implementing reservoir computing (RC) in hardware [32]. Reservoir computing requires nonlinear conversion of input to output signals along with a fading memory. Both features can be realized with OECT arrays. Further, variability in device characteristics due to the uncontrolled polymer microstructure can be exploited for realizing a multitude of relaxation timescales for the transistor On–Off characteristics. Using KCl electrolytic gates, the conductivity of the transistor channel can be increased under negative gate bias due to injection of negative ions. Upon removal of the bias, the ions diffuse back into the electrolyte and the channel recovers the original HRS. (Chapter 4 discussed the working principle in more detail.) Nonlinearity in the transistor transfer characteristics coupled with characteristic diffusional time constant for the ion transfer serve as building block properties for implementing a physical reservoir.

Operating voltage range was around −900 mV to avoid dissociation in the electrolyte while the short-term memory timescales were on the order of seconds. The variable response of the transistors in the array enables implicit representation of time for the RC. Signal classification was demonstrated with square wave and triangle-type pulses. Figure 5.9(a) shows the experiment protocol, and Figure 5.9(b) and (c) shows the current response of the 12 OECTs to square and triangle pulses, respectively. Figure 5.9(d) and (e) shows the response of the 12 transistors to the input patterns. Large D2D variability is seen in the current profile due to the varied polymer microstructure that is formed in the channel regions and is exploited for the signal processing. The output of the array is fed to a perceptron implemented in software for the classification. For this simple problem, the combination of hardware–software RC was able to classify signals with error rate approaching 0.001%. With such an approach, it becomes possible to integrate sensing (e.g., ion sensing) with computation (reservoir networks) in one platform.

Bio-compatible organic reservoir computing networks has been demonstrated with OECTs comprising poly(3,4-ethylenedioxythiophene) (PEDOT) doped with hexafluorophosphate (PF_6) [33]. In this work, dendritic connections between metal electrodes patterned on the semiconducting polymer devices were designed using tetrabutylammonium hexafluorophosphate ($TBAPF_6$) electrolyte medium. These connections act as reservoir networks with nonlinearity and short-term memory capabilities using similar principles of ionic displacement and accumulation at timescales faster than the relaxation time constants. The devices can be operated using PBS, an aqueous salt solution used to emulate that found in the human body. Classification of regular versus irregular heartbeats was demonstrated at an accuracy of 88% while flowers were classified at accuracy of 96% using standard datasets. Organic semiconductors therefore offer the exciting prospect of bridging neuromorphic computing with biocompatible and implantable electronics for real-time processing of signals generated in the body toward healthcare. It is also important to note that part of the analysis (linear regression) is performed in software. In future, it will be important to implement analysis such as matrix multiplication using hardware memory elements that are connected to the reservoir for data processing.

5.5 Spintronic Neural Networks

Following our discussions in Chapters 3 and 4 on electron spin-based neurons and synapses, we noted that they share very similar device structures. The neuron essentially is equivalent to a synaptic device interfaced with a reference MTJ. Figure 5.10 depicts an all-spin neural network organization where a cross-array of spin synapses is interfaced with spin neurons [34, 35]. RWL and WWL are the read and write control signals in the array which control the read and write operations of the synaptic devices respectively. Read operation for the synapse is equivalent to the write operation for the neuron MTJ. An appropriate magnitude of voltage applied along the BL lines can be used to either program the domain wall in the synapse (if WWL is activated) or is modulated by the MTJ conductance (if RWL is activated). Depending on the neuron device structure, we can map either an ANN or an SNN to this cross-array computing kernel.

5.5 Spintronic Neural Networks

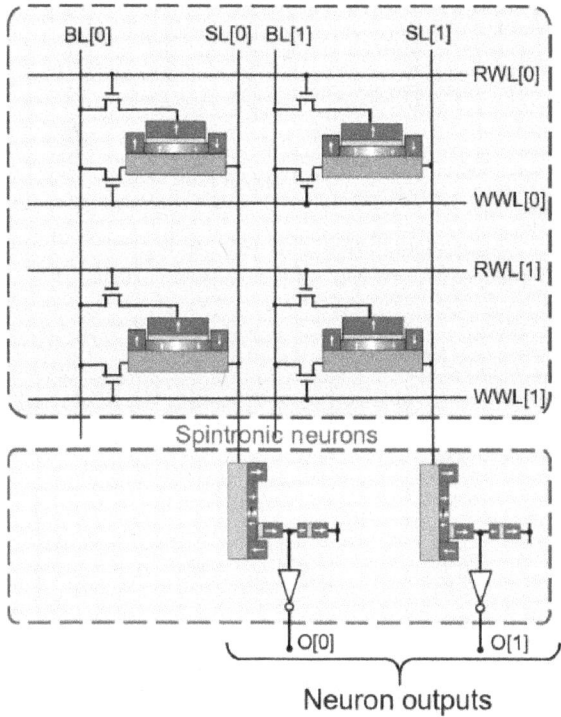

Figure 5.10 All-spin neural network where a cross-array of spin synapses drive spin neurons [39]. Adapted with permission from [39]. ©2020 IEEE.

So what are the specific device properties that will impact our circuit and system level design? The DW-MTJ neurons are magneto-metallic neurons requiring very low currents for switching (~µA). Due to the three-terminal device structure, this low current flows through the low-resistance HM layer and therefore this results in a low-voltage drop (~mV) across the spin neurons. The key property of low input resistance for the neuron along with the low current requirement is the main reason enabling the direct interfacing of spin synapse cross-arrays with spin neurons without the need for peripheral circuits like current-to-voltage converters (a key requirement in CMOS and other post-CMOS technology-based implementations [36, 37]). The low current requirement also enables us to drive the cross-array of synapses at low terminal voltages ~100 mV. This is critical from a system-level viewpoint. Standalone neuron energy consumption benefits may not necessarily translate to the system level, and therefore it is critical to consider codesign effects to evaluate the impact of peripheral circuitry and driving voltage requirements. Further, from a functionality standpoint, MTJ resistance is a function of applied voltage. If one can limit the applied voltage to levels ~100 mV, such resistance fluctuation effects can be minimized, thereby ensuring linear I–V characteristics

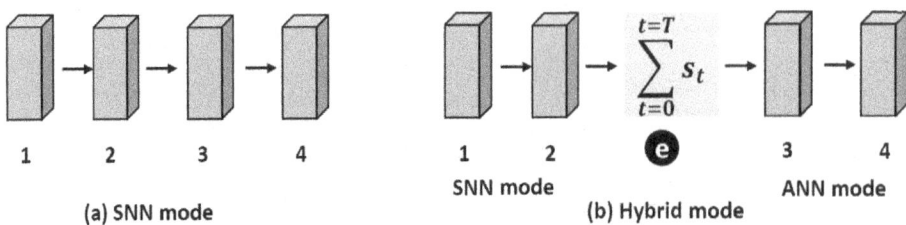

Figure 5.11 Conceptual depiction of (a) fully spiking and (b) hybrid ANN–SNN models [39]. Adapted with permission from [39]. ©2020 IEEE.

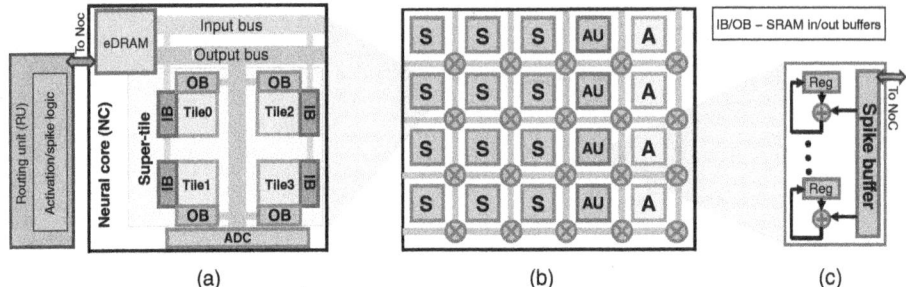

Figure 5.12 NEBULA architecture is depicted: (a) neural core (NC)s organized in a mesh NoC to implement (b) the hybrid ANN–SNN chip with (c) accumulator units (AU) serving as a bridge between ANN (A) and SNN (S) cores [39]. Adapted with permission from [39]. ©2020 IEEE.

during the synaptic read operation. Utilization of nonvolatile spin neurons is also critical to minimizing memory transactions for membrane potential state updates for a spiking neuron. System-level benchmarking has demonstrated the potential of ~100× improvement in energy consumption in comparison to iso-architecture CMOS implementations [34, 35].

Architectural explorations for all-spin neural networks have been investigated in literature [38, 39]. Noticing the device structure similarity between ANN and SNN neurons, Singh et al. [39] proposed a hybrid ANN–SNN architecture (see Figure 5.11) where the initial power-hungry layers of the network were operated in the SNN mode, while the low complexity deeper layers of the network were operated in the ANN mode. This helped to significantly reduce the inference latency for deep SNNs since spike propagation delay in deep SNNs can reduce its energy benefits. Figure 5.12 shows the architectural organization of NEBULA [39], which consists of predominantly SNN cores (S) along with ANN cores (A) interfaced with an accumulation unit (AU) to integrate spiking information into rate-coded inputs for the ANN. Each neuron core consists of an eDRAM memory for receiving network inputs along with SRAM input and output buffers (IB/OB).

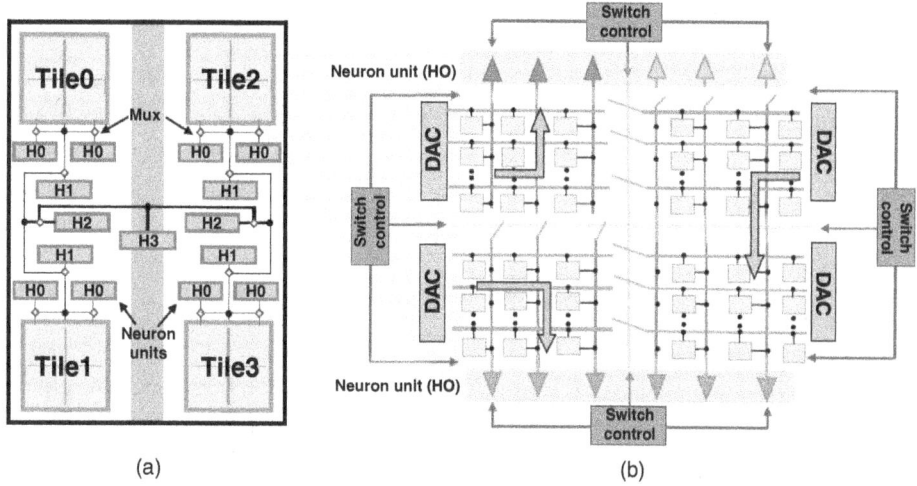

Figure 5.13 (a) A supertile consisting of 2 × 2 morphable tiles (see decomposition into ACs in (b)) and neuron unit hierarchy [39]. Adapted with permission from [39]. ©2020 IEEE.

The SRAM buffers supply inputs and store outputs from the cross-arrays. The cores are tiled in a mesh network on chip (NoC) architecture. Figure 5.13 shows the design of a neural core consisting of a super-tile of a 2 × 2 array of tiles (Tile0 to Tile3), which functions as the dot-product engine. In order to provide additional programmability in choosing the tile size (which impacts area utilization, energy, throughput, and peripheral cost), a morphable tile size design was considered.

The tile crossbar is further decomposable into a 2 × 2 array of smaller atomic crossbars (ACs) where the ACs can either function independently or together with other ACs in a tile, depending on the size of the mapped kernel. The neuron unit (NU) spin devices are attached along both the vertical extremities of the crossbar columns through switches. Further, spin-based devices provide additional benefits at the system level. In typical memristive technology-based architectural proposals [40–42], the dot-product operation is decomposed into partial products that are then mapped to multiple crossbars. Therefore, the analog crossbar outputs need to be first converted to digital signals using power-hungry Sense Amplifiers or ADCs in order to drive fan-out crossbars. For instance, a recent work reported that the power consumption of ADCs and DACs is around 72% of the total power consumption of their tile [40]. In order to minimize the high ADC overheads for aggregating partial sums, Singh et al. [39] considered a novel super-tile design with significantly reduced ADC usage. The central idea is to leverage the current-mode integrating property of the spin neurons to add the partial sums in the current domain itself by using Kirchhoff's Current Law and subsequently activating the appropriate NU hierarchy level depending on the size of the currently mapped kernel. The current-mode fusing is limited to adjacent tiles in a super-tile due to the challenges in current-mode signaling. ADC conversion takes place only when the data exits the super-tile. Benchmarking analysis has revealed that NEBULA has the potential to achieve 7.9× energy efficiency in the ANN mode and 45× energy efficiency in the SNN mode in comparison to state-of-the-art architectural designs for non-spin technologies [39].

5.6 Photonic Neural Networks

In Chapters 3 and 4, we have discussed possible photonic synapse and neuron implementations [43–46] based on optical switching of nonvolatile phase-change memory ($Ge_2Sb_2Te_5$ – GST) devices. In order to implement the parallel dot-product operation, a splitter can be employed to provide the wavelength division multiplexed (WDM) spiking signals to multiple rows of the photonic dot-product engine. The WDM signals (P_i s) are modulated by synaptic weights which are represented by the transmission coefficients (T_{ij} s). The outputs obtained from the photodetector array are then fed to laser diodes, which perform electrical current to optical spike conversion. Figure 5.14 shows a schematic of the dot-product engine.

Figure 5.15 depicts a detailed system schematic with the possibility of bipolar (positive and negative synaptic weights). Like conventional filamentary memristor systems, two photonic dot-product engines can be used to map the positive and negative weights. The positive and negative arrays drive the corresponding

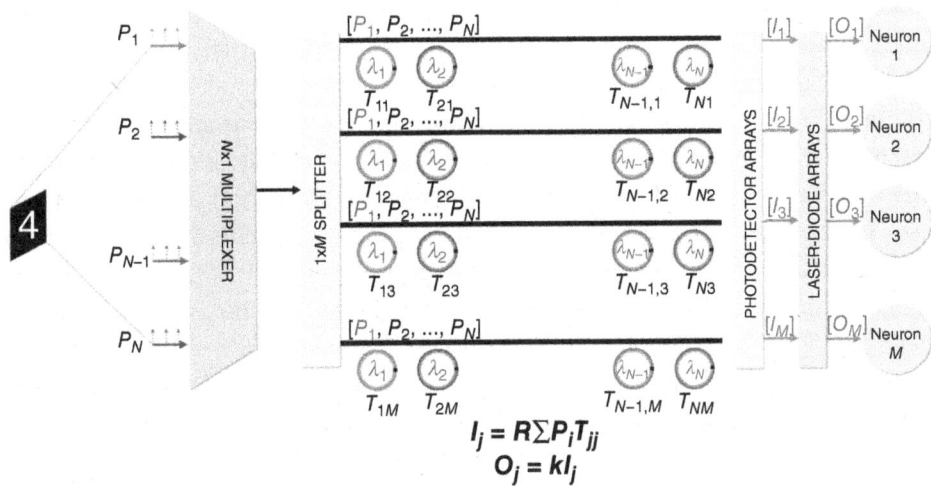

Figure 5.14 Optical dot-product engine consisting of an array of ring resonators with increasing radii which is interfaced with a photodetector and laser diode array [45]. Adapted with permission from [45]. © 2019 American Physical Society.

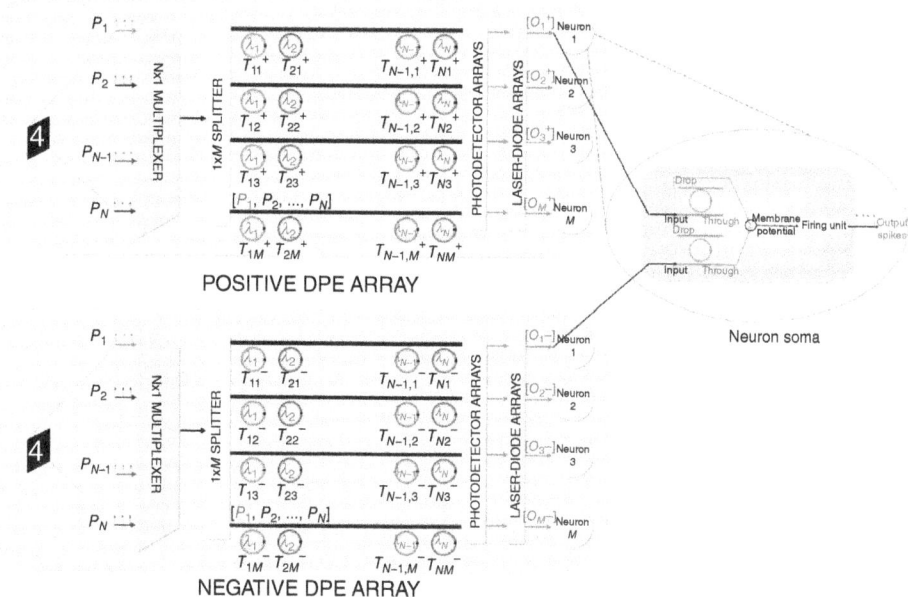

Figure 5.15 All-photonic neural network considering bipolar weights [45]. Adapted with permission from [45]. © 2019 American Physical Society.

positive and negative integrating ring resonators in the neurons. Since the two ring resonators integrate in opposite directions, bipolar synaptic integration can be achieved in the neuron membrane potential. As discussed before, the membrane potential unit is interfaced with a firing unit to determine if an output spike should be generated. The spike outputs can be used to drive fan-out photonic dot-product engines.

References

[1] Merolla, P. A., Arthur, J. V., Alvarez-Icaza, R., Cassidy, A. S., Sawada, J., Akopyan, F., Jackson, B. L., Imam, N., Guo, C., Nakamura, Y. and Brezzo, B., 2014. A million spiking-neuron integrated circuit with a scalable communication network and interface. *Science*, *345*(6197), pp. 668–673.

[2] Akopyan, F., Sawada, J., Cassidy, A., Alvarez-Icaza, R., Arthur, J., Merolla, P., Imam, N., Nakamura, Y., Datta, P., Nam, G. J. and Taba, B., 2015. Truenorth: Design and tool flow of a 65 mw 1 million neuron programmable neurosynaptic chip. *IEEE Transactions on Computer-Aided Design of Integrated Circuits and Systems*, *34*(10), pp. 1537–1557.

[3] Jeloka, S., Akesh, N. B., Sylvester, D. and Blaauw, D., 2016. A 28 nm configurable memory (TCAM/BCAM/SRAM) using push-rule 6T bit cell enabling logic-in-memory. *IEEE Journal of Solid-State Circuits*, *51*(4), pp. 1009–1021.

[4] Agrawal, A., Jaiswal, A., Lee, C. and Roy, K., 2018. X-SRAM: Enabling in-memory Boolean computations in CMOS static random access memories. *IEEE Transactions on Circuits and Systems I: Regular Papers*, *65*(12), pp. 4219–4232.

[5] Eckert, C., Wang, X., Wang, J., Subramaniyan, A., Iyer, R., Sylvester, D., Blaaauw, D. and Das, R., 2018, June. Neural cache: Bit-serial in-cache acceleration of deep neural networks. In *2018 ACM/IEEE 45th Annual International Symposium on Computer Architecture (ISCA)* (pp. 383–396). IEEE.

[6] Yin, S., Jiang, Z., Seo, J. S., and Seok, M., 2020. XNOR-SRAM: In-memory computing SRAM macro for binary/ternary deep neural networks. *IEEE Journal of Solid-State Circuits*, 55(6), pp. 1733–1743.

[7] Jaiswal, A., Chakraborty, I., Agrawal, A. and Roy, K., 2019. 8T SRAM cell as a multibit dot-product engine for beyond von Neumann computing. *IEEE Transactions on Very Large Scale Integration (VLSI) Systems*, 27(11), pp. 2556–2567.

[8] Kang, M., Lim, S., Gonugondla, S. and Shanbhag, N. R., 2018. An in-memory VLSI architecture for convolutional neural networks. *IEEE Journal on Emerging and Selected Topics in Circuits and Systems*, 8(3), pp. 494–505.

[9] Rathi, N., Chakraborty, I., Kosta, A., Sengupta, A., Ankit, A., Panda, P. and Roy, K., 2023. Exploring neuromorphic computing based on spiking neural networks: Algorithms to hardware. *ACM Computing Surveys*, 55(12), pp. 1–49.

[10] Akopyan, F., Sawada, J., Cassidy, A., Alvarez-Icaza, R., Arthur, J., Merolla, P., Imam, N., Nakamura, Y., Datta, P., Nam, G. J. and Taba, B., 2015. Truenorth: Design and tool flow of a 65 mw 1 million neuron programmable neurosynaptic chip. *IEEE Transactions on Computer-Aided Design of Integrated Circuits and Systems*, 34(10), pp. 1537–1557.

[11] Davies, M., Srinivasa, N., Lin, T. H., Chinya, G., Cao, Y., Choday, S. H., Dimou, G., Joshi, P., Imam, N., Jain, S. and Liao, Y., 2018. Loihi: A neuromorphic manycore processor with on-chip learning. *IEEE Micro*, 38(1), pp. 82–99.

[12] Boahen, K. A., 2000. Point-to-point connectivity between neuromorphic chips using address events. *IEEE Transactions on Circuits and Systems II: Analog and Digital Signal Processing*, 47(5), pp. 416–434.

[13] Moradi, S., Imam, N., Manohar, R., & Indiveri, G., 2013, September. A memory-efficient routing method for large-scale spiking neural networks. In *2013 European Conference on Circuit Theory and Design (ECCTD)* (pp. 1–4). IEEE.

[14] Carrillo, S., Harkin, J., McDaid, L. J., Pande, S., Cawley, S., McGinley, B. and Morgan, F., 2012, May. Hierarchical network-on-chip and traffic compression for spiking neural network implementations. In *2012 IEEE/ACM Sixth International Symposium on Networks-on-Chip* (pp. 83–90). IEEE.

[15] Davies, M., Wild, A., Orchard, G., Sandamirskaya, Y., Guerra, G. A. F., Joshi, P., Plank, P. and Risbud, S.R., 2021. Advancing neuromorphic computing with Loihi: A survey of results and outlook. *Proceedings of the IEEE*, 109(5), pp. 911–934.

[16] Deng, S., Yu, H., Park, T. J., Islam, A. N., Manna, S., Pofelski, A., Wang, Q., Zhu, Y., Sankaranarayanan, S. K., Sengupta, A. and Ramanathan, S., 2023. Selective area doping for Mott neuromorphic electronics. *Science Advances*, 9(11), p. eade4838.

[17] Lin, J., Guha, S. and Ramanathan, S., 2018. Vanadium dioxide circuits emulate neurological disorders. *Frontiers in Neuroscience*, 12, p. 856.

[18] Hoppensteadt, F. C. and Izhikevich, E. M., 1999. Oscillatory neurocomputers with dynamic connectivity. *Physical Review Letters*, 82(14), p. 2983.

[19] Delacour, C., Carapezzi, S., Abernot, M., Boschetto, G., Azemard, N., Salles, J., Gil, T. and Todri-Sanial, A., 2021, July. Oscillatory neural networks for edge AI computing. In *2021 IEEE Computer Society Annual Symposium on VLSI (ISVLSI)* (pp. 326–331). IEEE.

[20] Corti, E., Cornejo Jimenez, J. A., Niang, K. M., Robertson, J., Moselund, K. E., Gotsmann, B., Ionescu, A. M. and Karg, S., 2021. Coupled VO2 oscillators circuit as analog first layer filter in convolutional neural networks. *Frontiers in Neuroscience*, 15, p. 628254.

[21] Pickett, M. D., Medeiros-Ribeiro, G. and Williams, R. S., 2013. A scalable neuristor built with Mott memristors. *Nature Materials*, *12*(2), pp. 114–117.
[22] Han, C. Y., Fang, S. L., Cui, Y. L., Liu, W., Fan, S. Q., Huang, X. D., Li, X., Wang, X. L., Zhang, G. H., Tang, W. M. and Lai, P. T., 2023. Configurable NbOx memristors as artificial synapses or neurons achieved by regulating the forming compliance current for the spiking neural network. *Advanced Electronic Materials*, *9*(6), p. 2300018.
[23] Kim, H. W., Jeon, S., Kang, H., Hong, E., Kim, N. and Woo, J., 2023. Understanding rhythmic synchronization of oscillatory neural networks based on NbOx artificial neurons for edge detection. *IEEE Transactions on Electron Devices*, *70*(6), pp. 3031–3036.
[24] Delacour, C. and Todri-Sanial, A., 2021. Mapping Hebbian learning rules to coupling resistances for oscillatory neural networks. *Frontiers in Neuroscience*, *15*, p. 694549.
[25] Bo, Y., Zhang, P., Luo, Z., Li, S., Song, J. and Liu, X., 2020. NbO_2 memristive neurons for burst-based perceptron. *Advanced Intelligent Systems*, *2*(8), p. 2000066.
[26] Ha, S. D., Shi, J., Meroz, Y., Mahadevan, L. and Ramanathan, S., 2014. Neuromimetic circuits with synaptic devices based on strongly correlated electron systems. *Physical Review Applied*, *2*(6), p. 064003.
[27] Zhang, H. T., Park, T. J., Islam, A. N., Tran, D. S., Manna, S., Wang, Q., Mondal, S., Yu, H., Banik, S., Cheng, S. and Zhou, H., 2022. Reconfigurable perovskite nickelate electronics for artificial intelligence. *Science*, *375*(6580), pp. 533–539.
[28] Xia, Q., Robinett, W., Cumbie, M. W., Banerjee, N., Cardinali, T. J., Yang, J. J., Wu, W., Li, X., Tong, W. M., Strukov, D. B. and Snider, G. S., 2009. Memristor − CMOS hybrid integrated circuits for reconfigurable logic. *Nano Letters*, *9*(10), pp. 3640–3645.
[29] Camuñas-Mesa, L. A., Linares-Barranco, B. and Serrano-Gotarredona, T., 2019. Neuromorphic spiking neural networks and their memristor-CMOS hardware implementations. *Materials*, *12*(17), p. 2745.
[30] Krauhausen, I., Coen, C. T., Spolaor, S., Gkoupidenis, P. and van de Burgt, Y., 2024. Brain-inspired organic electronics: Merging neuromorphic computing and bioelectronics using conductive polymers. *Advanced Functional Materials*, *34*(15), p. 2307729.
[31] Demin, V. A., Erokhin, V. V., Emelyanov, A. V., Battistoni, S., Baldi, G., Iannotta, S., Kashkarov, P. K. and Kovalchuk, M. V., 2015. Hardware elementary perceptron based on polyaniline memristive devices. *Organic Electronics*, *25*, pp. 16–20.
[32] Pecqueur, S., Mastropasqua Talamo, M., Guérin, D., Blanchard, P., Roncali, J., Vuillaume, D. and Alibart, F., 2018. Neuromorphic time-dependent pattern classification with organic electrochemical transistor arrays. *Advanced Electronic Materials*, *4*(9), p. 1800166.
[33] Cucchi, M., Gruener, C., Petrauskas, L., Steiner, P., Tseng, H., Fischer, A., Penkovsky, B., Matthus, C., Birkholz, P., Kleemann, H. and Leo, K., 2021. Reservoir computing with biocompatible organic electrochemical networks for brain-inspired biosignal classification. *Science Advances*, *7*(34), p. eabh0693.
[34] Sengupta, A., Shim, Y. and Roy, K., 2016. Proposal for an all-spin artificial neural network: Emulating neural and synaptic functionalities through domain wall motion in ferromagnets. *IEEE Transactions on Biomedical Circuits and Systems*, *10*(6), pp. 1152–1160.
[35] Sengupta, A. and Roy, K., 2017. Encoding neural and synaptic functionalities in electron spin: A pathway to efficient neuromorphic computing. *Applied Physics Reviews*, *4*(4), pp. 041105-1–041105-23.

[36] Mochida, R., Kouno, K., Hayata, Y., Nakayama, M., Ono, T., Suwa, H., Yasuhara, R., Katayama, K., Mikawa, T. and Gohou, Y., 2018, June. A 4M synapses integrated analog ReRAM based 66.5 TOPS/W neural-network processor with cell current controlled writing and flexible network architecture. In *2018 IEEE Symposium on VLSI Technology* (pp. 175–176). IEEE.

[37] Xue, C. X., Chen, W. H., Liu, J. S., Li, J. F., Lin, W. Y., Lin, W. E., Wang, J. H., Wei, W. C., Chang, T. W., and Chang, M. F., 2019, February. 24.1 A 1Mb multibit ReRAM computing-in-memory macro with 14.6 ns parallel MAC computing time for CNN based AI edge processors. In *2019 IEEE International Solid-State Circuits Conference (ISSCC)* (pp. 388–390). IEEE.

[38] Sengupta, A., Ankit, A. and Roy, K., 2017, May. Performance analysis and benchmarking of all-spin spiking neural networks (special session paper). In *2017 International Joint Conference on Neural Networks (IJCNN)* (pp. 4557–4563). IEEE.

[39] Singh, S., Sarma, A., Jao, N., Pattnaik, A., Lu, S., Yang, K., Sengupta, A., Narayanan, V. and Das, C. R., 2020, May. Nebula: A neuromorphic spin-based ultra-low power architecture for SNNs and ANNs. In *2020 ACM/IEEE 47th annual international symposium on computer architecture (ISCA)* (pp. 363–376). IEEE.

[40] Shafiee, A., Nag, A., Muralimanohar, N., Balasubramonian, R., Strachan, J. P., Hu, M., Williams, R. S. and Srikumar, V., 2016. ISAAC: A convolutional neural network accelerator with in-situ analog arithmetic in crossbars. *ACM SIGARCH Computer Architecture News*, 44(3), pp. 14–26.

[41] Chi, P., Li, S., Xu, C., Zhang, T., Zhao, J., Liu, Y., Wang, Y. and Xie, Y., 2016. Prime: A novel processing-in-memory architecture for neural network computation in ReRAM-based main memory. *ACM SIGARCH Computer Architecture News*, 44(3), pp. 27–39.

[42] Song, L., Qian, X., Li, H. and Chen, Y., 2017, February. Pipelayer: A pipelined ReRAM-based accelerator for deep learning. In *2017 IEEE International Symposium on High Performance Computer Architecture (HPCA)* (pp. 541–552). IEEE.

[43] Ríos, C., Youngblood, N., Cheng, Z., Le Gallo, M., Pernice, W. H., Wright, C. D., Sebastian, A. and Bhaskaran, H., 2019. In-memory computing on a photonic platform. *Science Advances*, 5(2), p. eaau5759.

[44] Stegmaier, M., Ríos, C., Bhaskaran, H., Wright, C. D. and Pernice, W. H., 2017. Nonvolatile all-optical 1×2 switch for chipscale photonic networks. *Advanced Optical Materials*, 5(1), p. 1600346.

[45] Chakraborty, I., Saha, G. and Roy, K., 2019. Photonic in-memory computing primitive for spiking neural networks using phase-change materials. *Physical Review Applied*, 11(1), p. 014063.

[46] Chakraborty, I., Saha, G., Sengupta, A. and Roy, K., 2018. Toward fast neural computing using all-photonic phase change spiking neurons. *Scientific Reports*, 8(1), p. 12980.

6
System Design

6.1 Neuromorphic Architecture Design

Let us consider the computing primitive of an all-spin neural network [1] (see Figure 6.1) to illustrate the example of system design while considering optimization effects across the stack. We will begin with a brief refresher on the in-memory circuit and system operation. Each timestep of SNN operation corresponds to a spin neuron write–read–reset cycle. Each input is split up into two voltage lines V_{i+} (activated with a positive voltage upon spiking

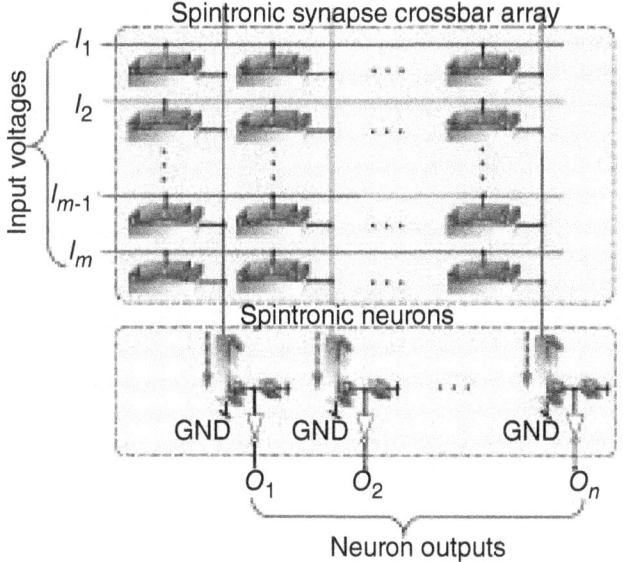

Figure 6.1 Core-computing primitive in an all-spin neural network [1]. The mapping is shown for a particular neural network layer with m inputs and n outputs. Adapted with permission from [1]. 2017 © AIP Publishing.

input) and V_{i-} (activated with a negative voltage upon spiking input) to implement bipolar weights. If the weight $w_{i,j}$ connecting the j-th neuron and i-th input is positive, then the spin device conductance $G_{i,j+}$ corresponding to V_{i+} is programmed to $G_{i,j+} = w_{i,j}G_0$, where G_0 is the mapped conductance for unity synaptic weight. The conductance $G_{i,j-}$ is programmed to a high OFF resistive state. Reverse mapping conditions are followed for negative weights. The crossbar columns are directly interfaced with the spin neurons where the output crossbar current drives the domain wall position in the spin neurons. The neuron device state is read during a subsequent read cycle and the device is reset if the neuron domain wall location reaches the opposite edge of the free layer in the magnet.

Let us consider the neuron input write resistance, that is, the input resistance of the heavy metal layer to be G_s and the voltage drop across the neuron to be V_s. By simple application of Kirchhoff's law and Ohm's law, it is possible to show, by equating the current supplied by the crossbar column to the current flowing through the neuron, that the net current supplied to the j-th spin neuron is,

$$I_j = G_s V_s = \frac{G_s \Sigma_i \left(G_{i,j+} V_{i+} + G_{i,j-} V_{i-} \right)}{G_s + \Sigma_i \left(G_{i,j+} + G_{i,j-} \right)} = \frac{\Sigma_i \left(G_{i,j+} V_{i+} + G_{i,j-} V_{i-} \right)}{1+\gamma}, \quad (6.1)$$

where $\gamma = \frac{\Sigma_i (G_{i,j+} + G_{i,j-})}{G_s}$ in Eq. (6.1) is a non-ideality factor in the dot-product calculation. Let us therefore consider this system next for understanding device–circuit–algorithm codesign principles.

6.2 Material and Device Design Impact on Circuit–Architecture Formulations

For this specific example, the control knobs for our design are the neuron and synapse device dimensions, crossbar supply voltage, crossbar resistance/conductance ranges, and spin neuron write latency. Note that we are not considering spin synapse write latency and other programming conditions since our example is corresponding to an inference operation and not learning. The system-level parameters that will be impacted by these control knobs are area, power, throughput, and energy consumption of our hardware along with the system-level accuracy.

From an algorithm standpoint, one needs to determine the minimum resolution necessary for synaptic and neuron state representation. Usually, higher precision is needed for synapses and more importantly for learning scenarios. However, for inference, both synapse and neuron state representations can be reduced to even single-bit levels leveraging recent algorithmic advancements at training binary neural networks [2]. The bit resolution will dictate the device length as higher synapse or neuron device length usually accommodates a greater number of available states as the domain wall can be programmed at multiple locations along the device. This is true even for other device technologies like ferroelectric field-effect transistors (FeFETs) where a larger device will be associated with a higher number of domains.

Next let us consider the impact of neuron device width. If we reduce the neuron device width, then the value of G_s reduces. This increases the non-ideality factor γ involved in Eq. (6.1) for the dot-product calculation which, in turn, would degrade the system-level accuracy. However, reducing the neuron device width is critical from a device scaling standpoint as this would reduce the current requirement from the crossbar arrays, thereby reducing the system-level energy consumption. However, the accuracy degradation can be potentially combatted by controlling the neuron device write latency. If we increase the write latency, we reduce the current requirement proportionately to drive the spin neurons for a given crossbar supply voltage. This would imply that the required range of synaptic conductances required to drive the spin neurons would reduce, effectively reducing the non-ideality factor γ. Such interplay considerations across the entire stack of materials and devices to circuits, systems, and algorithms are critical in neuromorphic system design.

6.3 Robust System Design Combatting Hardware Non-idealities

In-memory computing-based neuromorphic systems enabled by post-CMOS technologies suffer from intrinsic device non-idealities like cycle-to-cycle (C2C) and device-to-device (D2D) variations. A large section of works in reliable in-memory computing design has focused on expensive hardware-in-the-loop retraining or repeated evaluation-remapping methods [3–5]. In this chapter, we focus on a particular approach to combat non-idealities that avoids costly retraining by leveraging recent advances in Bayesian deep learning [6]. Bayesian deep networks is a particular type of neural network that are used for uncertainty quantification in confidence-critical applications like autonomous driving. In a Bayesian neural network, the network weights are conceptualized as probability distributions where the posterior is updated to closely match the prior distribution by minimizing the Kullback–Leibler (KL) divergence loss (measures the difference between two probability distributions) in addition to the normal loss function used in neural network training.

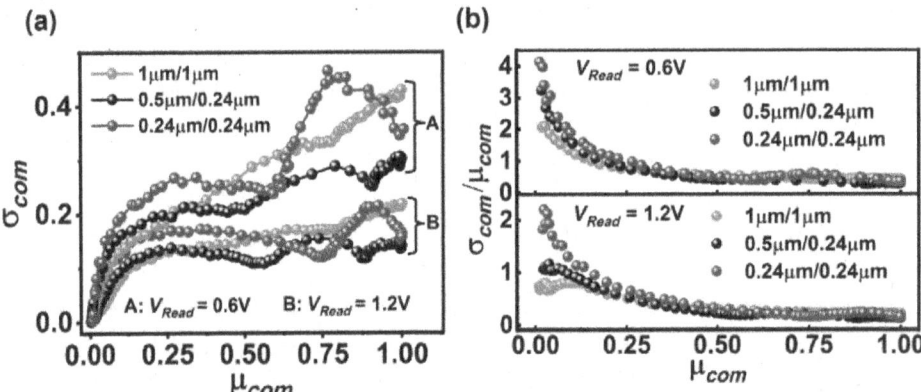

Figure 6.2 (a and b) Combined C2C and D2D variations (σ_{com}) as a function of the programmed state (μ_{com}) for a 28-nm high HKMG technology-based FeFET have been plotted for varying device sizes and read voltages [7]. Adapted with permission from [7]. Copyright © 2024, IEEE.

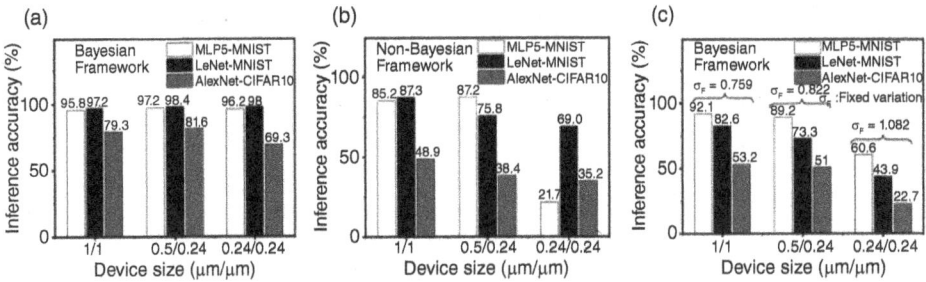

Figure 6.3 Inference accuracy for different neural network models corresponding to different device sizes at a read gate voltage of 0.6 V. Results are shown for training using the (a) Bayesian framework, (b) non-Bayesian framework, and (c) Bayesian framework where all network weights are subjected to a fixed amount of variation, σ_F [7]. Adapted with permission from [7]. Copyright © 2024, IEEE.

Figure 6.2 shows typical variation characteristics of a 28-nm high-k metal gate (HKMG)-based FeFET. As depicted in the figure, the combined variation profile is dependent on the device size and read voltage.

With reduction in device size, the variation profile experiences sharp kinks which is due to domain discretization effects of the underlying ferroelectric layer. The degree of variation also increases with reduction in the read gate voltage. Using the data for variations obtained from device characterization and modeling, such models can be considered as the intrinsic prior of a Bayesian neural network and the posterior can be trained to approximate this prior by minimizing the KL divergence between the two distributions. After training, the optimized mean values of the posterior distributions will serve as the weights to be programmed. The main advantage of this approach is that this does not require expensive hardware-in-the-loop retraining and essentially involves a one-time training process. As shown in Figure 6.3, the network trained using the Bayesian framework retains near-ideal accuracy under the given variation characteristics independent of device sizes [7]. We also note that considering the intrinsic device variability as prior is crucial in mitigating accuracy loss, as displayed by the data in Figure 6.3(c). It is worth mentioning here that the method is easily extensible to other emerging technologies as well.

6.4 Self-Healing Systems

While Section 6.3 discussed mitigation of variation effects in memristive devices, another significant concern is run-time non-idealities of the hardware which can include stuck-at-faults and resistance drift effects [8, 9]. Therefore, autonomous diagnosis and self-repair of the hardware is a highly desired functionality. Conventional approaches in this domain involve complex fault detection schemes [10, 11] and usually involve re-training using global backpropagation rules, thereby making it hardware inefficient. To address this issue, we take inspiration from astrocytes, a type of glial cell, that constitutes more than 50% of the cells in our brain and contributes to important functionalities such as self-repair and homeostatic regulation [12].

Figure 6.4 shows two neurons interacting with an astrocyte. Endocannabinoids mediated synaptic potentiation (e-SP) [13] and depolarization-induced suppression of excitation (DSE) [14] are two dynamic signals mediated by astrocytes

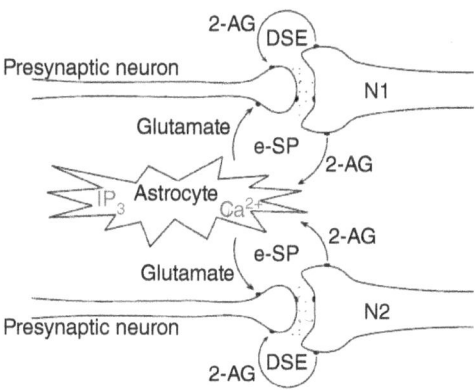

Figure 6.4 The astrocyte interacts as a common medium with the two neurons [9]. Adapted with permission from [9]. Copyright © 2023, Association for the Advancement of Artificial Intelligence (www.aaai.org). All rights reserved.

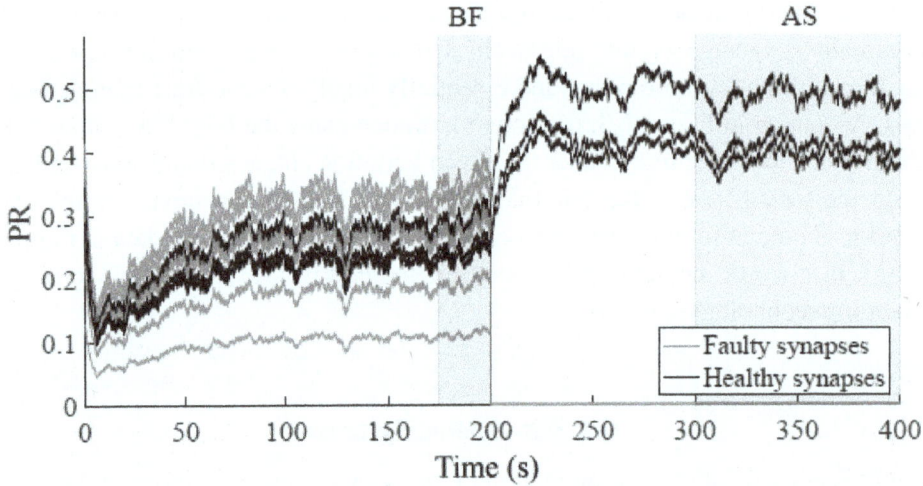

Figure 6.5 An example simulation of the temporal dynamics of the astrocyte–neuron network [9] where the PR values of synapses on neuron N2 are plotted. Ten synapses are considered, and fault occurs at 200s. Adapted with permission from [9]. Copyright © 2023, Association for the Advancement of Artificial Intelligence (www.aaai.org). All rights reserved.

that modulate the transmission probability (PR) of synapses. PR can be thought to be analogous to the synaptic weights that we consider in neural network training. Figure 6.5 depicts the temporal evolution of PR values of 10 synapses of neuron N2 of the neuron–astrocyte network considered in Figure 6.4. During the entire simulation window of 400 s, a fault is assumed to occur at 200 s. Upon fault injection, the PR of the faulty synapses decays to zero and the DSE of that corresponding neuron immediately reduces. However, the e-SP is a global signal and maintains its steady value due to continuous spiking activity of other neighboring neurons. This temporal fluctuation causes the PR of the healthy synapses to increase to improve the firing activity of the affected neuron to the baseline firing rate. Based on computational modeling outlined in Ref. [9], the PR values after fault are governed by Eq. (6.2):

$$PR_i(t) = q \times PR_i(BF) \left(1 - e^{-\frac{(t - t_{fault} + t_b)}{\tau}} \right), \qquad (6.2)$$

where $PR_i(t)$ is the PR value of the i-th synapse after fault at time t, $PR_i(BF)$ is the PR value before fault, q is the self-repair ratio [9] which is the ratio of the sum of the healthy synapses before and after fault, τ is the self-repair time constant, and t_{fault} is the timestep of fault occurrence. The temporal intercept t_b is given by Eq. (6.3):

$$t_b = -\tau \log\left(\frac{q-1}{q}\right). \tag{6.3}$$

Let us now consider embedding this astrocyte-modulated self-repair functionality in an unsupervised SNN network with STDP learning and lateral inhibition. The general STDP equations can be considered as:

$$\Delta w(t) = \begin{cases} \eta_{post} \times x_{pre}(t), & \Delta t > 0 \\ -\eta_{pre} \times x_{post}(t), & \Delta t < 0 \end{cases}, \tag{6.4}$$

where $\eta_{pre/post}$ are presynaptic/postsynaptic learning rates and $x_{pre/post}(t)$ are pre-synaptic/postsynaptic neuron traces, respectively. We note that the exponential dynamics mentioned in Eq. (6.2) can be represented by an equation of the form shown in Eq. (6.4), where the rate of weight change is directly proportional to the difference between the target synaptic PR and the current weight [9]:

$$\Delta w(t) = \begin{cases} \eta_{post} \times x_{pre}(t) \times \dfrac{qw_0 - w(t)}{\tau}, & \Delta t > 0 \\ -\eta_{pre} \times x_{post}(t), & \Delta t < 0 \end{cases}, \tag{6.5}$$

where $w(t)$ is the current weight and w_0 is the weight of the synapse before fault. It is important to note here that Eq. (6.5) is a local self-repair learning rule where the weight updates are dependent on the current weight (local to synapse) and sum of weights (local to neuron). The sum of weights is required to calculate the self-repair ratio q and can be easily tapped out of a crossbar array by passing high voltages across all rows of the array. We also do not require any fault-localization scheme as the self-repair process is autonomous and dependent on firing activities of the neurons. Extensive benchmarking analysis for various unsupervised vision tasks can be found in Ref. [9] where the astrocytic temporal regulation was found to significantly improve the final accuracy and repair speed in contrast to non-astrocyte based self-repair implementations.

6.5 Architecture–Application–Sensing Codesign

While Sections 6.2 and 6.3 have focused on codesign interplays across the stacks of devices, circuits, architectures, and algorithms, one must also be mindful of the way information is being sensed. For instance, a growing area of research in the neuromorphic computing community is the investigation of appropriate application drivers that may be catered for neuromorphic hardware and algorithms [15, 16]. The key notion is that applications characterized by a high degree of sparsity in input representations (spatially and/or temporally) may be an ideal fit for the

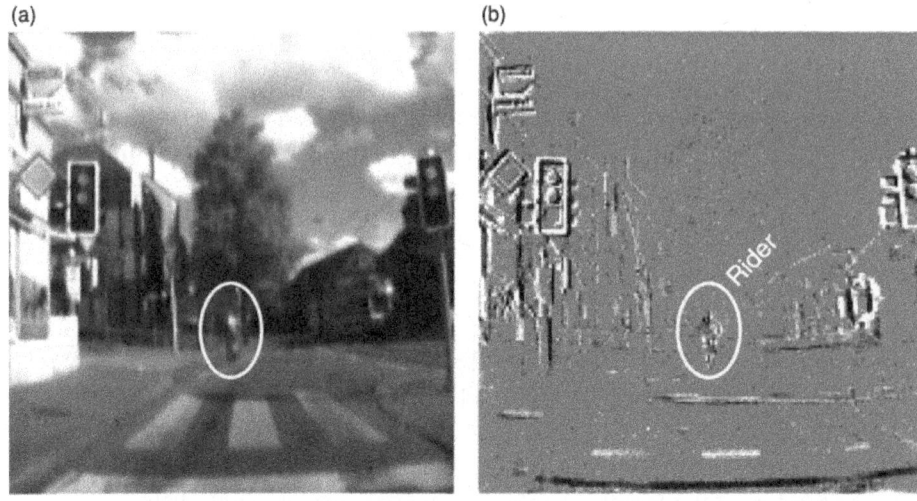

Figure 6.6 The output from a frame-based image sensor is shown in (a) while that from an event vision sensor is shown in (b). The output corresponds to a street scene recording from a moving car dashboard [18]. Adapted with permission from [18].

sparse, event-driven computing framework of SNN algorithms, thereby resulting in significant energy savings [16].

Let us now discuss a particular example in the context of dynamic vision sensor (DVS) cameras. Conventional cameras acquire information (absolute pixel intensities) at a specific frame rate. On the other hand, DVS cameras record changes in pixel intensity asynchronously, thereby resulting in fast- and low-power operation, and high dynamic range. Figure 6.6 depicts a typical output from a DVS camera. DVS cameras generate an output spike stream that encodes the time, location, and direction of pixel intensity change. Multiple works have explored SNN algorithms for event-sensor applications [15, 17]. One solution in the domain of ANN–SNN conversion is where the input events from the camera are accumulated to construct event-frames which can be used to train an ANN and subsequently converted to an SNN. However, the sparse temporal data obtained from event-sensors are usually associated with considerable noise which can result in accuracy drops. To mitigate such effects, Ref. [15] considers noise filtering strategies like combining multiple frames and operating the first layer as an ANN. Alternatively, other works have also explored hybrid ANN–SNN networks to deal with such effects [17]. While algorithm designs are being revisited for such applications, it is important to note that such hybrid algorithm designs are intrinsically tied to the architectural layer (for instance, hybrid ANN–SNN design needs to consider different hardware computing cores and data communication styles), thereby requiring an across the stack optimization.

References

[1] Sengupta, A. and Roy, K., 2017. Encoding neural and synaptic functionalities in electron spin: A pathway to efficient neuromorphic computing. *Applied Physics Reviews*, 4(4), pp. 041105-1–041105-23.

[2] Rastegari, M., Ordonez, V., Redmon, J. and Farhadi, A., 2016, September. Xnor-net: Imagenet classification using binary convolutional neural networks. In *European conference on computer vision* (pp. 525–542). Cham: Springer International Publishing.

[3] Liu, B., Li, H., Chen, Y., Li, X., Wu, Q. and Huang, T., 2015, June. Vortex: Variation-aware training for memristor X-bar. In *Proceedings of the 52nd Annual Design Automation Conference* (pp. 1–6).

[4] Chen, L., Li, J., Chen, Y., Deng, Q., Shen, J., Liang, X. and Jiang, L., 2017, March. Accelerator-friendly neural-network training: Learning variations and defects in RRAM crossbar. In *Design, Automation & Test in Europe Conference & Exhibition (DATE), 2017* (pp. 19–24). IEEE.

[5] Jin, S., Pei, S. and Wang, Y., 2020, March. On improving fault tolerance of memristor crossbar based neural network designs by target sparsifying. In *2020 Design, Automation & Test in Europe Conference & Exhibition (DATE)* (pp. 91–96). IEEE.

[6] Wang, H. and Yeung, D. Y., 2020. A survey on Bayesian deep learning. *ACM Computing Surveys (CSUR)*, 53(5), pp. 1–37.

[7] Manna, B., Saha, A., Jiang, Z., Ni, K. and Sengupta, A., 2024. Variation-resilient FeFET-based in-memory computing leveraging probabilistic deep learning. *IEEE Transactions on Electron Devices*, 71(5), pp. 2963–2969.

[8] Ambrogio, S., Gallot, M., Spoon, K., Tsai, H., Mackin, C., Wesson, M., Kariyappa, S., Narayanan, P., Liu, C. C., Kumar, A., Chen, A., and Burr, G. W., 2019, December. Reducing the impact of phase-change memory conductance drift on the inference of large-scale hardware neural networks. In *2019 IEEE International Electron Devices Meeting (IEDM)* (pp. 6–1). IEEE.

[9] Han, Z., Islam, A. N. and Sengupta, A., 2023, June. Astromorphic self-repair of neuromorphic hardware systems. In *Proceedings of the AAAI conference on artificial intelligence* (Vol. 37, No. 6, pp. 7821–7829).

[10] Chen, T. J., Li, J. F. and Tseng, T. W., 2012. Cost-efficient built-in redundancy analysis with optimal repair rate for RAMs. *IEEE Transactions on Computer-Aided Design of Integrated Circuits and Systems*, 31(6), pp. 930–940.

[11] Xia, L., Liu, M., Ning, X., Chakrabarty, K. and Wang, Y., 2017, June. Fault-tolerant training with on-line fault detection for RRAM-based neural computing systems. In *Proceedings of the 54th Annual Design Automation Conference 2017* (pp. 1–6).

[12] Oberheim, N. A., Wang, X., Goldman, S. and Nedergaard, M., 2006. Astrocytic complexity distinguishes the human brain. *Trends in Neurosciences*, 29(10), pp. 547–553.

[13] Navarrete, M. and Araque, A., 2010. Endocannabinoids potentiate synaptic transmission through stimulation of astrocytes. *Neuron*, 68(1), pp. 113–126.

[14] Alger, B. E., 2002. Retrograde signaling in the regulation of synaptic transmission: Focus on endocannabinoids. *Progress in Neurobiology*, 68(4), pp. 247–286.

[15] Singh, S., Sarma, A., Lu, S., Sengupta, A., Narayanan, V. and Das, C. R., 2021, July. Gesture-SNN: Co-optimizing accuracy, latency and energy of SNNs for neuromorphic vision sensors. In *2021 IEEE/ACM International Symposium on Low Power Electronics and Design (ISLPED)* (pp. 1–6). IEEE.

[16] Davies, M., Wild, A., Orchard, G., Sandamirskaya, Y., Guerra, G. A. F., Joshi, P., Plank, P. and Risbud, S. R., 2021. Advancing neuromorphic computing with Loihi: A survey of results and outlook. *Proceedings of the IEEE*, 109(5), pp. 911–934.

[17] Lee, C., Kosta, A. K., Zhu, A. Z., Chaney, K., Daniilidis, K. and Roy, K., 2020, August. Spike-flownet: Event-based optical flow estimation with energy-efficient hybrid neural networks. In *European conference on computer vision* (pp. 366–382). Cham: Springer International Publishing.

[18] Brandli, C., Berner, R., Yang, M., Liu, S. C. and Delbruck, T., 2014. A 240 × 180 130 db 3 µs latency global shutter spatiotemporal vision sensor. *IEEE Journal of Solid-State Circuits*, *49*(10), pp. 2333–2341.

7

Neuromorphic Algorithms

7.1 Spiking Neural Network Training Methodologies

In this chapter, we will focus on spiking neural network (SNN) learning algorithms that consider the discrete, temporal nature of spikes during model training formulation. While the event-driven, sparse spiking behavior is a critical feature of SNNs that enables its power efficiency, it presents unique challenges for algorithmic researchers since they are a paradigm shift in the context of current deep learning algorithms. Figure 7.1 provides a summary of common SNN training algorithms in the domain of unsupervised and supervised learning.

Let us first consider Figure 7.1(a) which depicts an unsupervised mechanism of weight update in SNNs termed as spike-timing-dependent plasticity (STDP). STDP has been a traditional SNN training paradigm inspired from winner-take-all (WTA) networks in cortical microcircuits of the brain [1]. Considering spike information transmitted from a pre-neuron to a post-neuron to be modulated by the synaptic efficacy (w), STDP weight update can be formulated as a function of the time of spike of the pre-neuron (t_{pre}) and post-neuron (t_{post}):

$$\Delta w = \begin{cases} A_+ \exp\left(\dfrac{-\Delta t}{\tau_+}\right), & \Delta t > 0 \\ -A_- \exp\left(\dfrac{\Delta t}{\tau_-}\right), & \Delta t < 0 \end{cases}, \quad (7.1)$$

where $\Delta t = t_{post} - t_{pre}$, and $A_+, A_-, \tau_+,$ and τ_- are constants. The characteristics of Equation 7.1 are based on experimental measurements performed in rat hippocampal glutamatergic synapses [1]. The essence of this unsupervised learning rule is that neurons that fire together wire together. As is evident from the equation, STDP weight update promotes temporal correlation/de-correlation effect. If the post-neuron fires after (before) the pre-neuron, then we can assume that

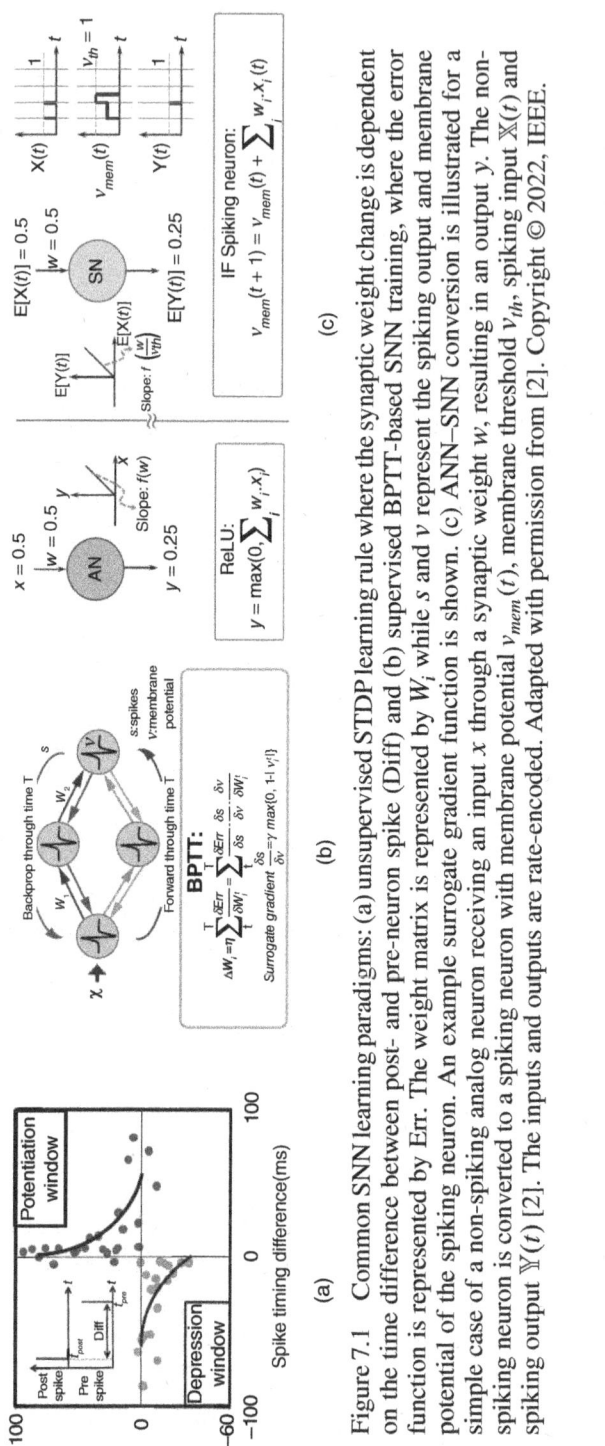

Figure 7.1 Common SNN learning paradigms: (a) unsupervised STDP learning rule where the synaptic weight change is dependent on the time difference between post- and pre-neuron spike (Diff) and (b) supervised BPTT-based SNN training, where the error function is represented by Err. The weight matrix is represented by W_i while s and v represent the spiking output and membrane potential of the spiking neuron. An example surrogate gradient function is shown. (c) ANN–SNN conversion is illustrated for a simple case of a non-spiking analog neuron receiving an input x through a synaptic weight w, resulting in an output y. The non-spiking neuron is converted to a spiking neuron with membrane potential $v_{mem}(t)$, membrane threshold v_{th}, spiking input $\mathbb{X}(t)$ and spiking output $\mathbb{Y}(t)$ [2]. The inputs and outputs are rate-encoded. Adapted with permission from [2]. Copyright © 2022, IEEE.

7.1 Spiking Neural Network Training Methodologies

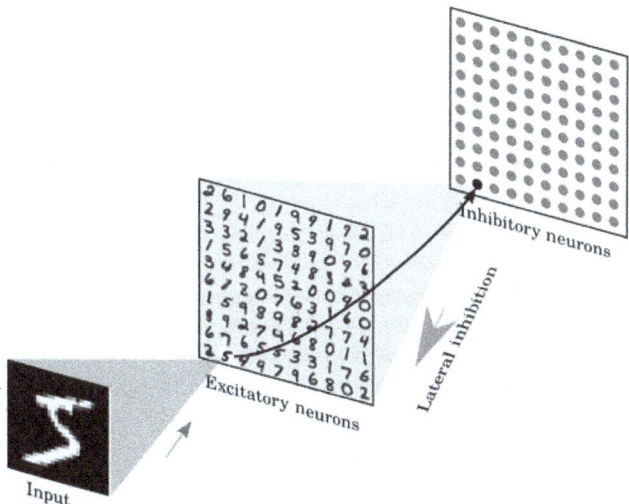

Figure 7.2 WTA network architecture where neurons in the excitatory layer learn representative patterns from the input spike data using STDP. The excitatory neurons are equipped with lateral inhibitory neurons that promote competitiveness [5]. Adapted with permission from [5] under the Creative Commons Attribution 4.0 license.

the post-neuron firing is related (unrelated) to the pre-neuron spike, and therefore the corresponding synaptic weight should be increased (decreased). The change in synaptic strength is exponentially related to the spike timing difference. From an edge computing perspective, the key advantages of STDP-based learning rule is that it is local, event-driven, and unsupervised as one is no longer reliant on global backpropagation-based model updates that require huge amount of labeled data. It is worth mentioning that while the STDP rule shown in Figure 7.1(a) is an antisymmetric one, other variants like symmetric STDP-based weight updates [3] have been also used in literature and have significance for specific applications like associative computing [4].

Next, consider the WTA network [5] shown in Figure 7.2 which consists of an excitatory layer of neurons with lateral inhibitory connections. The lateral inhibitory effect is implemented through an inhibitory layer of neurons, each of which receives a direct connection from the corresponding excitatory neuron and is recurrently connected to all the other neurons in the excitatory layer through inhibitory (negative) synapses. Lateral inhibition promotes competitive learning in the WTA network. The input spike train triggers the neurons in the excitatory layer and the weight map joining the inputs and the excitatory neurons are learnt using STDP. Each neuron in the excitatory layer is also equipped with homeostasis effect, where the neuron membrane potential threshold increases every time the neuron fires, making it more difficult for the neuron to fire in the future. This

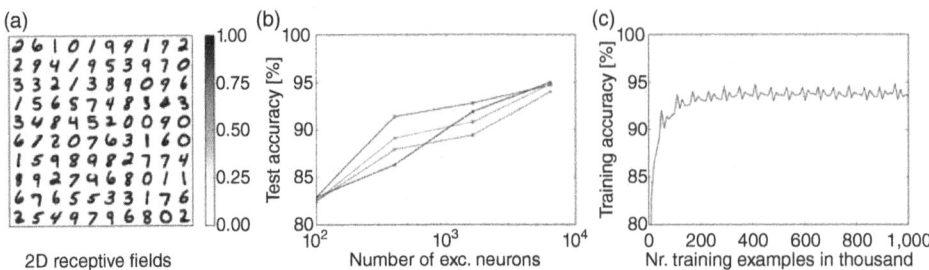

Figure 7.3 (a) Learnt weights of the 100-neuron network plotted in 28 × 28 fashion similar to image dimensions from the MNIST dataset. The neurons learn generic representations of the input images. (b) The test accuracy can be increased by increasing the number of excitatory neurons. (c) Training accuracy saturates with only a subset of the entire training set [5]. Adapted with permission from [5] under the Creative Commons Attribution 4.0 license.

provides equal opportunity to all neurons in the excitatory later to fire and prevents single neurons from dominating the firing pattern due to the lateral inhibitory effect. Once the learning is completed, a small subset of the training set can be passed through the network to assign each excitatory neuron to a corresponding label by noting its maximum firing activity. This neuron label can be used subsequently during the testing phase by observing the maximum firing activity of neurons in the excitatory pool in response to an input.

Figure 7.3(a) shows representative weights learnt by a 100-neuron network in response to MNIST digit images rate-encoded as spike trains. The test accuracy can be increased by increasing the number of neurons (Figure 7.3(b)), while the network accuracy saturates upon being presented with a subset of the training set (Figure 7.3(c)). This is a critical advantageous feature of such networks where learning can be performed with a limited amount of labeled data. While accuracies exhibited by STDP trained networks are lower than their supervised deep learning counterparts, it is worth noting that this training methodology is specifically catered for edge computing based on-chip learning scenario where a human annotator will not be available to provide labeled data in real time. Research efforts have also been directed to train deep convolutional architectures using such learning rules, although with limited benefits [6].

In the domain of supervised learning, back-propagation through time (BPTT) (Figure 7.1(b)) has been used to train SNNs using similar concepts borrowed from recurrent neural network (RNN) training [7, 8]. Due to the temporal input integration property, SNNs can be inherently recurrent in nature. The SNN is unrolled over time, and then BPTT is applied to update the weights:

$$\Delta w_{ij} = -\alpha \sum_t \frac{\partial L}{\partial w_{ij}} = -\alpha \sum_t \frac{\partial L}{\partial o_i^t} \frac{\partial o_i^t}{\partial u_i^t} \frac{\partial u_i^t}{\partial w_{ij}} = -\alpha \sum_t \frac{\partial L}{\partial o_i^t} \sigma'\left(u_i^t\right) \frac{\partial u_i^t}{\partial w_{ij}}, \quad (7.2)$$

7.1 Spiking Neural Network Training Methodologies

where α is the learning rate and L is the loss function. w_{ij} represents the weight connecting neurons i and j. o_i^t and u_i^t represent the spiking output and membrane potential of neuron i at timestep t. Since the spiking activations of neurons in SNNs are discrete events, surrogate gradients are used to deal with the non-differentiability issue of spiking neurons. $\sigma'(u_i^t)$ represents the surrogate function used for approximating the gradient. An example surrogate function is shown in Figure 7.1(b) and further examples can be found in References [9, 10].

While BPTT-based SNN training is a promising route to train large-scale ANNs, it suffers from significant memory usage due to temporal unrolling as well as spike non-differentiability issues. As a near-term alternative, ANN–SNN conversion has been proposed where an analog neural network (ANN) is first trained using standard principles of backpropagation for non-spiking networks and is then subsequently converted to a spiking network for event-driven inference. ANN–SNN conversion relies on the fact that an integrate-fire spiking neuron without any leak and refractory period bears functional equivalence to a ReLU neuron [11]. This has been shown conceptually using a particular example in Figure 7.1(c). The transfer function of the spiking neuron is represented by the total number of output spikes versus the total number of input spikes. The positive half of the transfer function corresponds to the case of a positive weight providing the input (in the simple case of a singular input as shown here). The rate of output firing frequency increases linearly with the rate of input firing frequency, thereby mimicking ReLU functionality. However, for the case of negative weight, the membrane potential of the spiking neuron never increases to cross the threshold membrane potential, thereby producing zero spikes. The slope of the transfer curve is a function of the ratio of the weights of the trained ANN to the spiking threshold in the converted SNN. A high spiking threshold can result in increased inference latency, which becomes a significant problem for deep SNNs. On the other hand, a low choice of spiking threshold can result in incorrect spiking neuron operation as the information to be accumulated to the membrane potential can potentially lie above the firing threshold. Therefore, a proper normalization strategy of the ratio of weights to threshold is necessary to ensure that the threshold of the spiking neurons is sufficiently high to prevent incorrect inference without unnecessarily increasing the latency [12]. Other architectural constraints like avoiding bias, batch normalization, and max-pooling layers may also be necessary [12]. For instance, including a bias term expands the threshold balancing optimization landscape to include an additional factor per layer, thereby making the process more complex. Avoiding bias would necessitate non-utilization of batch normalization during training as that introduces a bias-like term during the inference process. Therefore, dropout can be used as an alternative regularizer during the training process. Further, the binary nature of spiking output is not

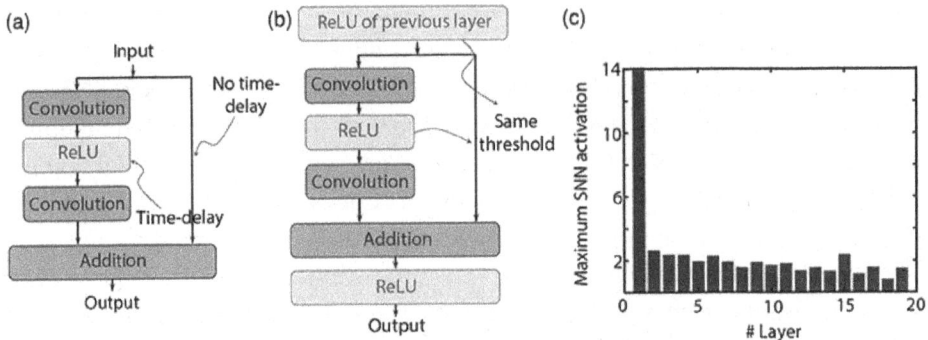

Figure 7.4 (a) Basic functional unit of ResNet. (b) Design constraints included in the ResNet functional unit to ensure near-lossless conversion [12]. (c) The bar chart shows the maximum activation of neurons in SNN with the increase in the number of layers. Adapted with permission from [12] under the Creative Commons Attribution 4.0 license.

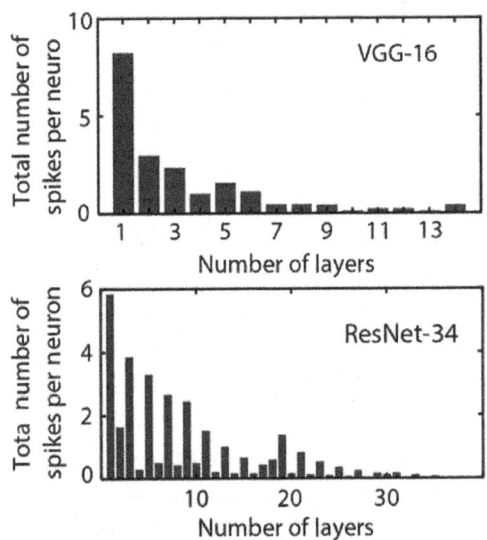

Figure 7.5 Spiking sparsity of the converted SNN increases significantly with increasing layer depth for VGG and residual architectures on the ImageNet dataset [12] under the Creative Commons Attribution 4.0 license.

compatible with max-pooling operator. Therefore, ANNs catered for conversion to SNNs are usually trained with average pooling.

Specific design constraints can also be required for different neural architectures. For instance, consider residual networks (ResNet) as shown in Figure 7.4. The basic functional unit of a ResNet is shown in Figure 7.4(a), which results in unbalanced temporal delay in the two parallel paths. A solution would be to include ReLUs at the two junction points (Figure 7.4(b)), which when converted to spiking neurons, would result in intrinsic temporal delays, thereby assisting in balancing the delays on the two parallel paths. Further, the two fan-in ReLU layers to the bottom ReLU layer need to be converted to spiking layers with the same threshold in order to ensure that similar analog outputs from the two layers are rated encoded with same frequency [12]. The advantage of event-driven inference is apparent in Figure 7.5, where the average spiking rate (ASR) of individual layers reduces significantly as layer depth increases, thereby resulting in up to an order of magnitude energy reduction [12].

7.2 Local Learning

Although BPTT-based SNN training has shown recent success for deep networks, especially using hybrid techniques [13, 14] (to be discussed in Section 7.6), it is still considered to be biologically implausible. BPTT assumes symmetric weights in forward and backward propagation along with nonlocal weight updates based on gradient calculation and error backpropagation, which are in stark contrast to local learning mechanisms observed in the brain [15–17]. In this section, we will review two additional SNN learning mechanisms with varying degrees of locality (see Figure 7.6) [8].

We have already discussed global BPTT-based SNN training and ANN–SNN conversion in Section 7.1. Feedback alignment relaxes the symmetric weight connection constraint in BPTT by introducing random matrices in the backward pass. E-prop is a variant of direct feedback alignment [18] where the error calculated at the output layer is propagated directly to each layer instead of layer-wise propagation for weight calculation. DECOLLE [19] is a more localized learning rule where a loss function is calculated at each layer by using a random matrix to map the layer output to pseudo-targets. This method also avoids layer-wise error backpropagation. We will discuss the two methods, namely, e-prop and DECOLLE in more detail next.

The e-prop learning mechanism [18] is based on the concept of synaptic eligibility trace [20] which can be calculated as:

$$e_{ij}^t = \frac{\partial o_i^t}{\partial w_{ij}}. \tag{7.3}$$

The eligibility trace measures the impact of synaptic weight on the temporal spiking activity of the postsynaptic neuron. Along with the eligibility trace, a learning signal z_i^t corresponding to the i-th neuron at timestep t is used for weight update as follows:

$$\Delta w_{ij} = -\alpha \sum_t z_i^t \cdot e_{ij}^t. \tag{7.4}$$

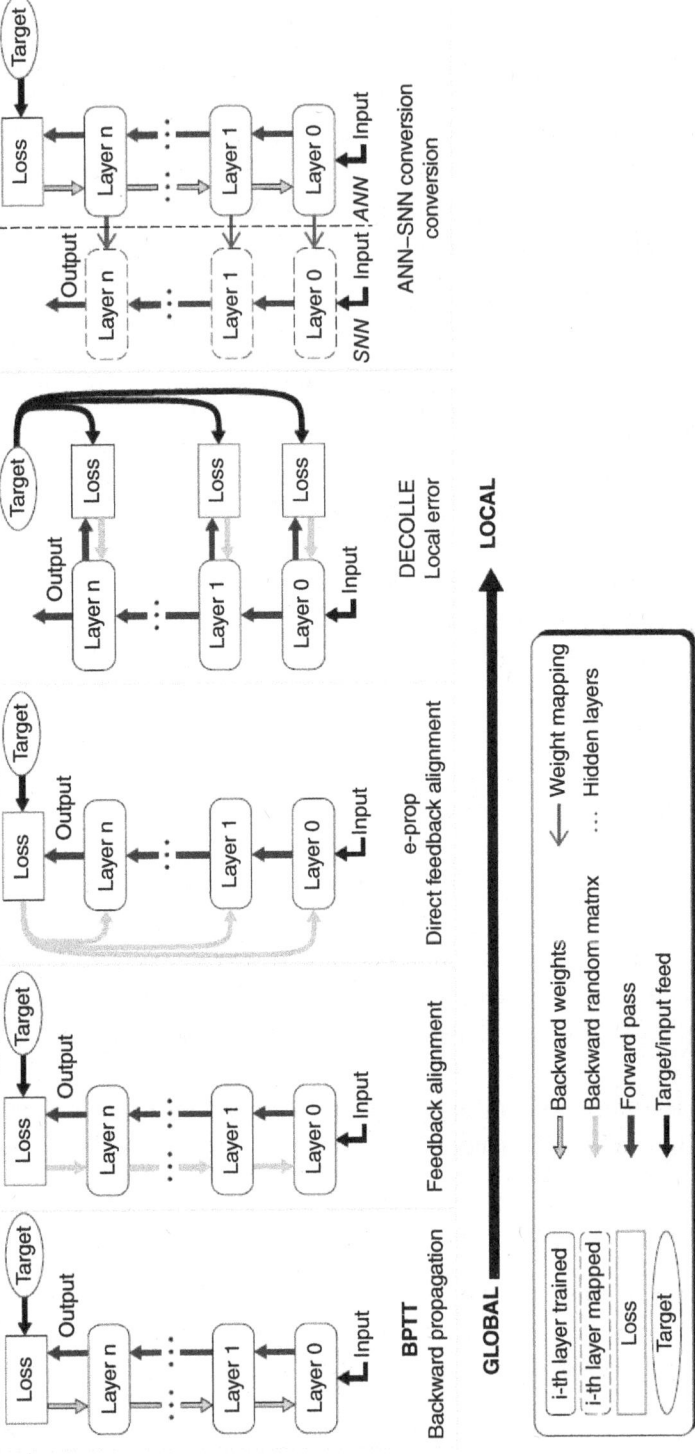

Figure 7.6 Different SNN training methodologies are shown in increasing degree of locality: BPTT, feedback alignment, e-prop, and DECOLLE. ANN–SNN conversion is also shown on the far right [8]. Adapted with permission from [8].

Based on direct feedback alignment, the learning signal is computed by using a random weight matrix g_{ik} to scale the derivative of the loss function at the output layer:

$$z_i^t = \sum_k g_{ik} \frac{\partial L_k}{\partial o_k^t}. \tag{7.5}$$

As mentioned earlier, the error signal is propagated directly from the output layer to intermediate layers, thereby avoiding global backpropagation of errors as seen in BPTT-based training.

On the other hand, Deep Continuous Local Learning (DECOLLE) [19] uses random readouts at each layer, $y_i^{l,t}$ corresponding to layer l, and timestep t through linear projection of random matrices g_{ij}^l:

$$y_i^{l,t} = \sum_j g_{ij}^l o_j^{l,t}. \tag{7.6}$$

The loss L^t is calculated as the sum of the layer-wise differences between these readouts and the pseudo-targets. The local weight update equations for a particular layer, where all nonlocal gradients corresponding to other layers are set to zero, are given as follows:

$$\Delta w_{ij}^t = -\alpha \frac{\partial L^t}{\partial w_{ij}^t} = -\alpha \frac{\partial L^t}{\partial o_i^t} \frac{\partial o_i^t}{\partial w_{ij}^t} \tag{7.7}$$

$$\frac{\partial o_i^t}{\partial w_{ij}^t} = \sigma'(u_i^t) \left(p_i^t - \rho \frac{\partial r_i^t}{\partial w_{ij}^t} \right). \tag{7.8}$$

Here, p_i^t denotes the membrane potential trace of the i-th neuron driven by incoming spikes. The refractory dynamics is encoded in r_i^t along with the constant factor ρ. Considering that the refractory dynamics has negligible impact on the membrane potential, the weight update equation can be simplified as follows:

$$\Delta w_{ij}^t = -\alpha \frac{\partial L^t}{\partial o_i^t} \sigma'(u_i^t) p_i^t. \tag{7.9}$$

While each of these algorithms are being thoroughly investigated to overcome their respective shortcomings and scaling challenges, an initial benchmarking effort can be found in Reference [8]. Local learning rules usually have some amount of accuracy degradation. Further, recurrent SNN architectures exhibit enhanced accuracies.

7.3 Probabilistic Computing

While we have primarily discussed SNN algorithms with deterministic neurons so far, a growing body of neuromorphic algorithmic works have focused on stochastic neurons, inspired by the probabilistic firing of biological neurons [23].

While deterministic neurons are closer to standard deep learning formulations from the algorithm side, stochastic neurons solve the non-differentiability issue of deterministic spiking neurons [10] while providing fault tolerance [24]. On the other hand, the neuromorphic hardware community has been actively exploring stochastic neuronal and synaptic devices from a scaling perspective. As these devices are scaled down, they are no longer able to support multiple nonvolatile programming states. In addition, they are characterized by enhanced stochasticity (cycle-to-cycle variations) during the switching process. Therefore, research has started in earnest to investigate whether the information encoded in multi-bit deterministic states can be translated to the information encoded in stochastic switching of a binary element [25]. For instance, a neural activation of 0.4 that lies between 0 and 1 can be represented by a single bit element characterized by 40% switching probability.

Networks with stochastic sigmoid neurons in the domain of rate encoding can be either trained directly [26] or through simple conversion techniques [27] following principles we have already discussed in Section 7.1 for deterministic neurons. Probabilistic versions of multi-bit synaptic plasticity have also been explored where the state change is represented by the switching probability of the synapse [25, 28]. Recent works [29–31] have also attempted to combine stochastic spiking neural networks with temporal encoding techniques where the decision is based on precise timing of spikes. Temporal encoding reduces the inference latency and enhances the spiking sparsity of SNNs significantly in contrast to rate encoding techniques where information is transmitted in the firing frequency of neurons over a large time window. Let us now discuss the algorithmic formulation for stochastic SNNs trained with temporal encoding [30]. Similar formulations can be extended to deterministic neurons as well.

Consider a stochastic SNN where the neuron's membrane potential at timestep t is represented by u_i^t. The neuron is triggered by spikes o_j^t through synaptic weights w_{ij}. The probability of firing of the stochastic neuron at the t-th timestep can be given by:

$$p_i^t = \sigma\left(u_i^t\right) = \sigma\left(\frac{\lambda u_i^{t-1} + \sum_j w_{ij} o_j^t}{k_i}\right), \tag{7.10}$$

where σ stands for the sigmoid nonlinearity in Equation 7.10 and k_i is a scaling factor. λ represents the impact of leak in the membrane potential dynamics. Based on first-to-spike coding, we calculate the probability P_t of the correct output neuron to produce the earliest spike at timestep t:

$$P_t = p_c^t \prod_{i=1, i \neq c}^{n} \prod_{t'=1}^{t} \left(1 - p_i^{t'}\right) \prod_{t'=1}^{t-1} \left(1 - p_c^{t'}\right), \tag{7.11}$$

where p_c^t denotes the probability of the correct neuron c producing a spike at the t-th timestep. Equation 7.11 is based on the concept that no wrong neurons should generate a spike before the correct neuron. It is worth noting here that Bernoulli function can be used to determine the spike values generated by each neuron from their corresponding spike probabilities. Straight-through estimator (STE) [32] can be used to overcome the non-differentiable nature of the Bernoulli function. The maximum likelihood criterion [29] can be used to calculate the total loss function L as follows:

$$L = \log\left(\sum_t P_t\right). \tag{7.12}$$

As per Equation 7.12, P_t reduces with increase in timesteps, which in turn reduces the overall loss function. Therefore, this encourages the neurons to fire earlier but not before the correct neuron. Stochastic SNNs with temporal encoding are characterized by reduced latencies in contrast to their deterministic counterparts with temporal encoding [30]. The BPTT algorithm can be used to minimize the loss function mentioned in Equation 7.12 by time unrolling the network. Further benchmarking results exploring design trade-offs in terms of accuracy, sparsity, latency, and energy efficiency can be found in Reference [30].

7.4 Dynamic Networks

A large section of our discussion so far has focused on training with stationary data distributions. However, in lifelong learning scenarios, a neural network is exposed to new data incrementally. Current AI systems are unable to cope with such incremental learning environments, resulting in catastrophic forgetting of previously learnt tasks while trying to learn the new task [33, 34]. However, the human brain is highly plastic and involves the formation of new neurons (neurogenesis) and synapses (synaptogenesis) that represent an integral part of nervous system development [35]. Neurogenesis in adult brains has been considered to be essential for cognitive adaptability [36] and brain injury recovery [37].

Inspired by the concept of neurogenesis, recent efforts have considered dynamic neural networks [38] to cope with the issue of catastrophic forgetting where the neural network architecture adapts dynamically to changing data distributions. This enables the network to efficiently manage resources while providing better accuracies than iso-architecture static networks [39]. In this context, we will discuss Grow When Required (GWR) networks which creates or removes neurons and synapses based on principles of Hebbian learning [35, 40]. A conceptual depiction of GWR network operation is shown in Figure 7.7.

Consider the example of an Hebbian learning-trained SNN (discussed in Section 7.1) which performs unsupervised clustering on the input data. When presented with a new input $x(t)$, GWR networks monitor the spiking response of the winning neuron. If the spiking response of the winning neuron is below a threshold, the network adds a new node to account for the novel input. The weight map of the new node is given by:

$$w_{new} = \frac{w_o + x(t)}{2}. \tag{7.13}$$

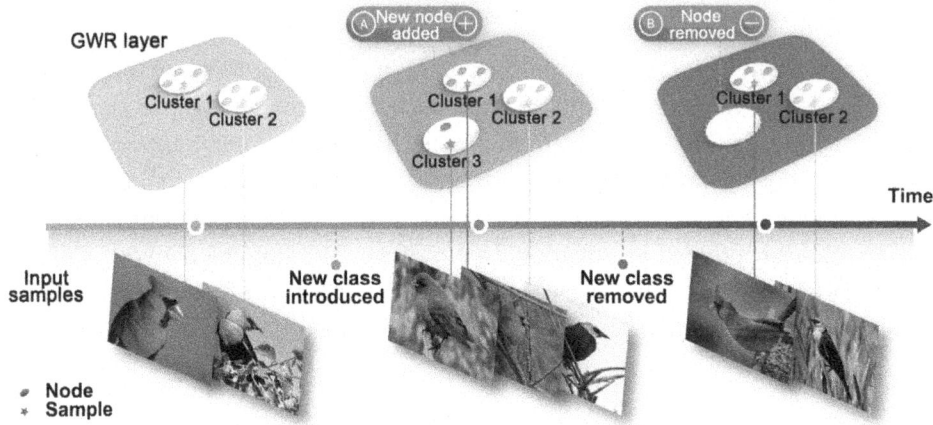

Grow when required (GWR) network

Figure 7.7 Conceptual depiction of GWR networks where the network performs unsupervised clustering on the input data. As the network is presented with novel inputs, it adds new nodes to learn the new class. Conversely, the network removes redundant nodes when not presented with particular classes over time [35]. Adapted with permission from [35]. Copyright © 2022, The American Association for the Advancement of Science.

Figure 7.8 (a) An incremental learning scenario is considered in this example by presenting the neural network with 10,000 examples from the first five classes (0–4) of the MNIST dataset. Subsequently, the network is trained on 20,000 examples from all classes of the dataset. The network grows to account for the novel data without suffering accuracy loss (shown by the bar chart in the top right corner). Finally, we introduce a final training phase where we present the network with examples only from classes 0 to 4. The network shrinks by removing redundant nodes. (b–e) Test accuracy and number of nodes of a static and dynamic network in an incremental learning scenario for the MNIST and CUB-200 datasets [35]. (f–i) The associated learnt weight representations of the network are shown. Adapted with permission from [35]. Copyright © 2022, The American Association for the Advancement of Science.

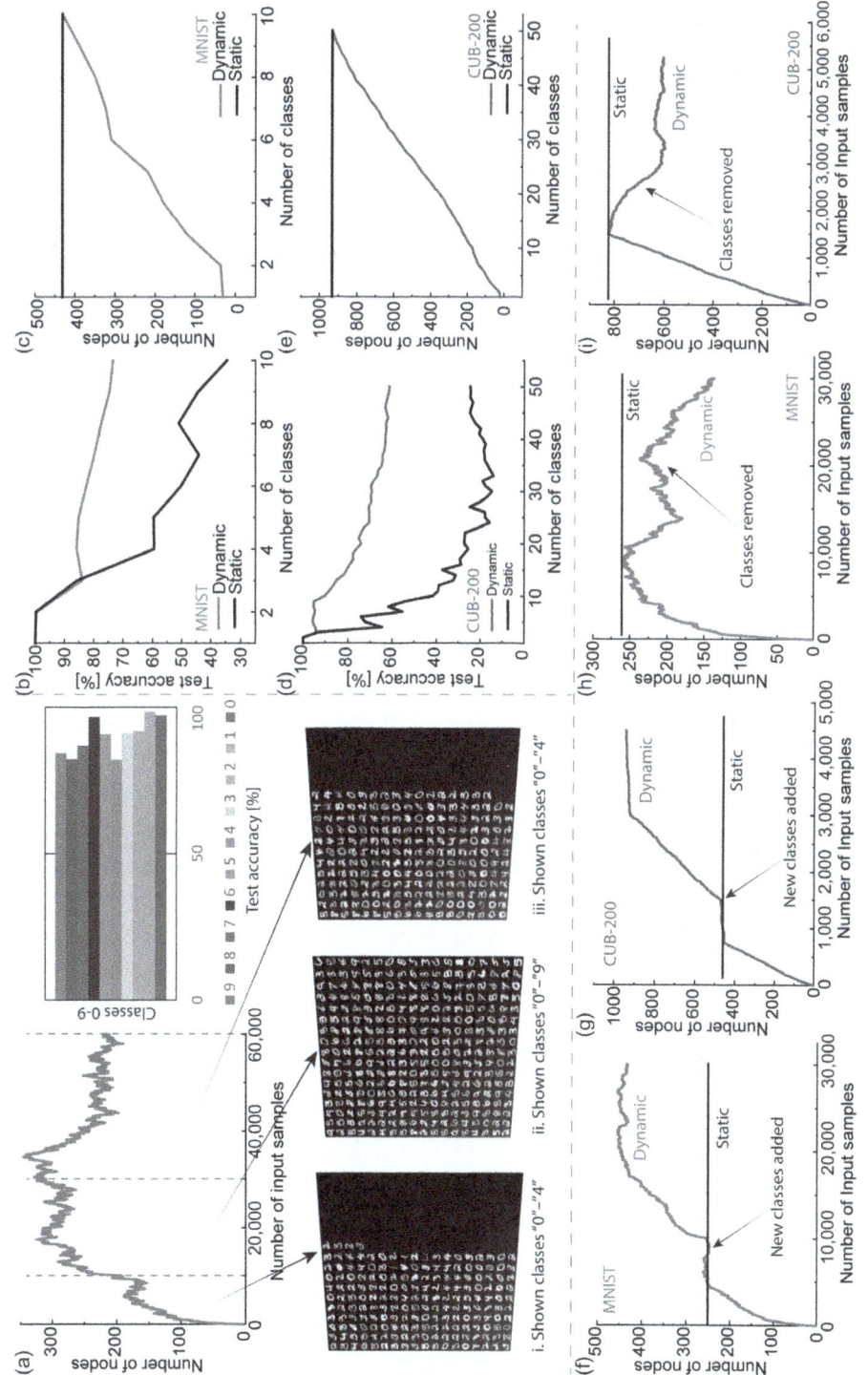

Here, w_o is the weight map of the winning neuron. GWR networks prioritizes learning of new nodes over old ones by scaling their weight updates by an activity factor β_i, which is initialized to 1 and decays exponentially over time. If a node has not fired for a long time, the corresponding nodes are removed. The general weight update equation of the GWR network is given by (in the absence of node addition process):

$$\Delta w_i = \alpha \beta_i \left(x(t) - w_i \right). \tag{7.14}$$

Here, α represents the learning rate. This weight update scheme is applied selectively to the winning node and the second-best winning neuron to ensure that the neurons best approximating the input data are nudged toward the input data distribution provided its spiking activity is above a particular threshold. For more information on training algorithm formulation, interested readers are referred to Reference [40]. Performance benefits of considering dynamic networks in AI hardware development for lifelong learning was recently explored in Reference [35] (see Figure 7.8).

7.5 Energy-Based Models and Learning

Spiking neural network-based algorithm explorations have been primarily limited to static vision-based tasks and convolutional architectures [41]. In the era of large language models (LLMs), it is critical to address scalability challenges of SNNs – for instance, incorporation of attention mechanisms and applicability to sequential tasks. BPTT-based SNN training is not scalable to LLM architectures since it stores the entire computational graph for training and requires time-based unrolling. To address these challenges, we consider a complementary perspective of rethinking SNNs as event-driven dynamical systems [41]. The concept that neurons collectively adjust their states in response to external stimulus to better explain the input data has been a popular hypothesis [42, 43]. We will discuss two techniques in this context – equilibrium propagation (EP) [44] and implicit differentiation at equilibrium (IDE) [45].

Equilibrium propagation is an energy-based learning algorithm formulation for convergent recurrent neural network (RNN) architectures with static input at each timestep. Specifically, the weights between two layers of the network are bidirectional and symmetric. The network dynamics follows two stages, namely, a "free" phase and a "nudge" phase. In the "free" phase, the network dynamics is governed by the equation:

$$s_{t+1} = \frac{\partial \phi}{\partial s}(x, s_t, \theta), \tag{7.15}$$

7.5 Energy-Based Models and Learning

where s_t represents the combined (including all the layers) state of the RNN at timestep t, x represents the input and θ denotes the model parameters. The scalar primitive function that dictates the transition dynamics is represented by ϕ. The "free" phase concludes with the convergence of the network dynamics to a steady-state s^*. The state transition dynamics of the subsequent "nudge" phase is given by:

$$s_{t+1}^{\beta} = \frac{\partial \phi}{\partial s}\left(x, s_t^{\beta}, \theta\right) - \beta \frac{\partial L}{\partial s}\left(s_t^{\beta}, y\right), \tag{7.16}$$

where the additional term $-\beta \frac{\partial L}{\partial s}$ causes the state of the output layer to be nudged in the direction of loss function (L) minimization. β serves as a scaling factor and y is the target. Perturbations from the output layer causes the hidden layers to converge to the weakly clamped fixed point s_β^*. Following the two phases, the weight update equation of the network can be written as:

$$\nabla_{EP}^{\theta}(\beta) = \frac{1}{\beta}\left(\frac{\partial \phi}{\partial \theta}\left(x, s_\beta^*, \theta\right) - \frac{\partial \phi}{\partial \theta}\left(x, s^*, \theta\right)\right). \tag{7.17}$$

If the scalar primitive function is derived from a Hopfield-like energy function, it can be shown that the weight updates, shown in Equation 7.17, can be computed locally in space [44] and time [46]. It is worth pointing out here that EP, unlike BPTT, uses the same computational circuitry during both phases of training [44]. While EP has strong theoretical connections with recurrent BP [47], it exhibits STDP-like weight updates [1, 44], thereby being a potential fit for neuromorphic hardware with local on-chip learning facilities exhibiting orders of magnitude energy savings [48].

Recent work [49] has explored application of EP in sequence learning tasks by merging the EP framework with a self-attention mechanism implemented using modern Hopfield networks. Modern Hopfield networks [50, 51] is also an energy-based model where the state transition dynamics is given by:

$$\xi^{new} = X \, softmax\left(\beta_h X^T \xi\right), \tag{7.18}$$

where ξ^{new} is the retrieved pattern, X is the stored pattern, and ξ is the state pattern. Given the constant β_h governing the state transition dynamics, the network converges to rich metastable states where the retrieved state is like the self-attention mechanism found in transformers (a type of neural network architecture that includes self-attention [49] mechanism and is the backbone behind the recent popularity of LLMs. It is worth observing the similarity between Equation 7.18 and self-attention equation [49]). While both EP and modern Hopfield networks are energy-based formulations, it should be noted here that the state transition dynamics of modern Hopfield networks have single step convergence guarantee [50] and therefore helps in maintaining the convergence of EP. The conceptual overview of

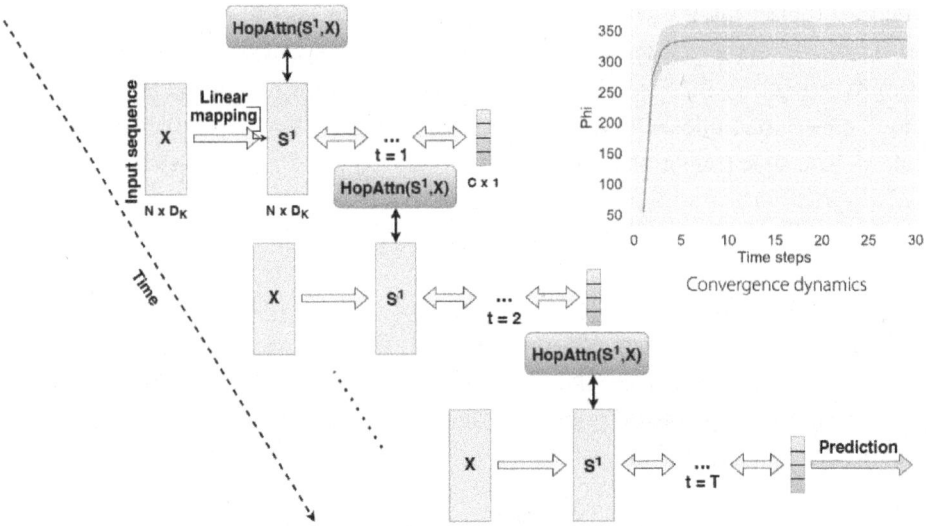

Figure 7.9 Conceptual overview of the operation of a convergent RNN integrated with Hopfield attention module trained using EP during the "free" phase. The convergence of the scalar primitive function with time is shown in the upper right [41]. Adapted with permission from [41]. Copyright © 2024, IEEE.

the integration of the two models and the convergence of the scalar primitive function is shown in Figure 7.9. The state transition dynamics of the i-th layer where Hopfield attention (*HopAttn*) is applied is given by:

$$s_{t+1}^i = HopAttn\left(\frac{\partial \phi}{\partial s^i}(x, s_t, \theta), X\right), \qquad (7.19)$$

where the input X to the neural network is equivalent to the sequence of stored patterns. State-of-the-art performance of such attention integrated EP-trained networks on sequence learning tasks can be found in Reference [49]. Extension of spiking EP-trained networks to convolutional architectures have been also explored [52].

Let us now shift the focus from local learning to scaling SNN architectures in the domain of natural language processing (NLP). We leverage the concept of implicit differentiation that has been used in deep equilibrium models [53]. Conceptualizing the SNN as a dynamical system, IDE leverages the steady-state convergence property of the ASR of neurons in the SNN to train the model parameters instead of time-based unrolling used in BPTT, thereby avoiding scalability challenges.

Consider a fixed-point equation of the form $z = f_\theta(z)$, where θ represents the model parameters. The fixed-point equation converges over time, ultimately reaching an equilibrium state. The equation can be alternatively posed as, $g_\theta(z) = f_\theta(z) - z$, where $g_\theta(z)$ converges to 0 at the equilibrium state z^*.

Using the loss function at equilibrium $L(z^*)$, the following relation can be derived using IDE [53]:

$$\frac{\partial L(z^*)}{\partial \theta} = -\frac{\partial L(z^*)}{\partial z^*}\left(J_{g_\theta}^{-1}\Big|_{z^*}\right)\frac{\partial f_\theta(z^*)}{\partial \theta}, \qquad (7.20)$$

where $J_{g_\theta}^{-1}\big|_{z^*}$ is the inverse Jacobian of g_θ at equilibrium. As shown in Equation 7.20, the calculated gradients are dependent only on the final converged state, thereby avoiding the need to store intermediate neuron states as well as encountering issues associated with spiking discontinuities [54]. Let us now discuss this in the context of SNN training. In the discrete time setting, the membrane potential dynamics of a spiking neuron is given by:

$$u_i[t+1] = \gamma u_i[t] + W_{(i-1)} \cdot s_{(i-1)}[t+1] + b_i - V_{th}s_i[t+1], \qquad (7.21)$$

where $u_i[t]$ is the membrane potential of neurons in layer i at timestep t, $s_i[t]$ are the spikes generated by layer i at timestep t, γ is the leak term, W_i is the operation performed by the i-th layer on the input spikes, b_i is a bias term, and V_{th} is the spiking threshold. The ASR of neurons in the i-th layer can be defined as:

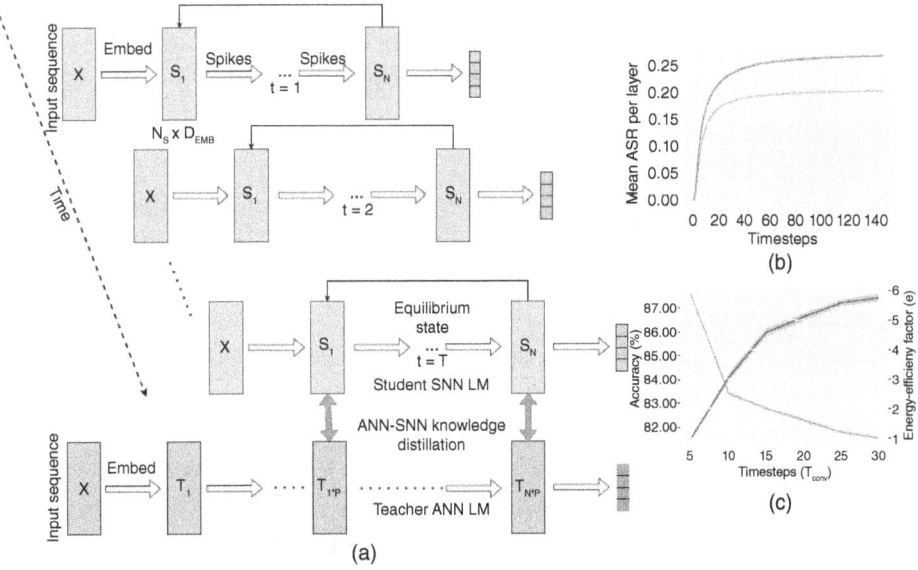

Figure 7.10 (a) IDE-based SNN training where weight updates are based on the final converged ASR values of the various layers. The steady-state convergence also enables us to utilize knowledge distillation between a non-spiking "teacher" and a spiking "student" model. (b) Convergence of mean ASR values of two different layers in the SNN architecture. (c) Accuracy and energy-efficiency trade-off of the SNN LLM with respect to timesteps [41]. Adapted with permission from [41]. Copyright © 2024, IEEE.

$$a_i[t] = \frac{\sum_{\tau=1}^{t} \gamma^{t-\tau} s_i[\tau]}{\sum_{\tau=1}^{t} \gamma^{t-\tau}}. \tag{7.22}$$

With time, the ASR of various layers in the SNN converges toward equilibrium values a_i^* and satisfies the following relationship:

$$a_i^* = \sigma\left(\frac{W_{(i-1)} a_{(i-1)}^* + b_i}{V_{th}}\right), \tag{7.23}$$

where $\sigma(x)$ is a clipping function constraining the values between 0 and 1. While Equation 7.23 can be easily derived for linear layers [45], Reference [54] empirically demonstrates convergence for nonlinear layers as well. The steady-state equations for the layers also enables us to utilize an ANN–SNN Knowledge Distillation framework (see Figure 7.10) where the loss function is measured between the intermediate layer outputs of a trained "teacher" model and ASR of the corresponding layers in the "student" SNN model. Recent works have explored extreme quantization in such IDE-enabled spiking LLMs by incorporating the quantization process in the training framework [55].

7.6 Hybrid Algorithm Designs Leveraging Conventional Deep Learning Aspects

While snapshots of recent ongoing work in the domain of brain-inspired neuromorphic algorithms have been provided earlier, it is important to highlight that a large amount of research in the community also points toward hybrid algorithm and system design combining conventional deep learning aspects [56]. For instance, STDP training can be used as an incremental training strategy for adapting to online changing environments. Reference [57] has considered STDP as a pretraining strategy for faster training convergence and improved robustness. In the domain of supervised learning, some approaches [13] have attempted to combine ANN–SNN conversion and BPTT-based SNN training in order to reduce the computational overhead of BPTT training. Even gradient-free optimizers like neuroevolution [58] have been used to intelligently adapt spiking thresholds of the SNN, thereby completely avoiding the BPTT-based fine-tuning stage after conversion [59]. Application- and hardware-level feedback is also critical for designing such hybrid neuromorphic systems. For instance, Reference [60] considers a hybrid ANN–SNN system design where the power-hungry initial layers of the network are operated in SNN mode and the later layers are operated in the ANN mode to reduce the overall inference latency. Such a hybrid system design was evaluated to be 1.1–2.5× energy efficient in comparison to a pure

SNN design. Multimodal pattern recognition systems like Tianjic [61] have been also designed where ANN modules are used for object detection and tracking and SNN is used for voice recognition.

7.7 Need for Hardware–Algorithm–Application–Neuroscience Codesign

Chapters 5 and 6 have already outlined the critical need for hardware–algorithm–application codesign. However, while event-driven spiking neural systems are an initial step to expanding the impact of neuroscience on AI system design, it is more important than ever to forge stronger connections with neuroscience in the design of brain-inspired algorithms and hardware (Figure 7.11) [62].

We have previously discussed aspects related to novel components in the brain like astrocytes or glial cells in Chapter 6 for self-repair functionalities of AI hardware [63]. Research is also underway to explore other complementary perspectives like dendritic computing [64, 65]. Incorporation of architectural organization aspects inspired by the brain should also be considered. For instance, Bayesian learning approaches that combine top-down and bottom-up information flow (updating posterior probability distribution based on observed data in addition to feedforward information propagation) might be critical to design AI systems that can respond to uncertain sensory data [66]. In summary, while success of deep

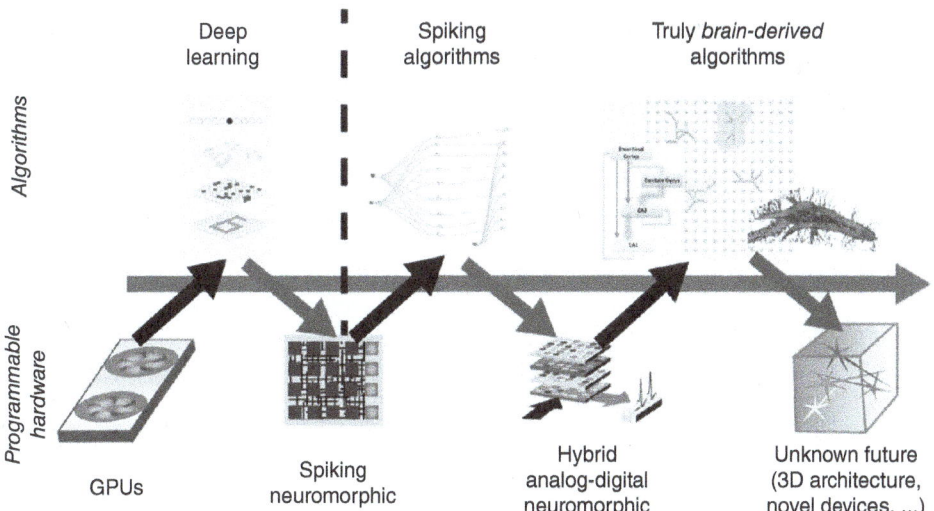

Figure 7.11 Roadmap for brain-inspired computing driven by hardware–algorithm–application–neuroscience codesign perspective [62]. Adapted with permission from [62]. © 2020 The Authors. Advanced Intelligent Systems published by Wiley-VCH GmbH.

The technical gap between neuroscience and computing has limited brain inspiration in technology.

The success of artificial neural networks has enabled a *narrow* transfer of knowledge from neuroscience.

Neuromorphic technologies can broadly reduce the gap between computing and neuroscience, leading to a greater knowledge transfer.

Figure 7.12 Neuromorphic computing has potential to significantly bridge the knowledge gap between the neuroscience and computing communities [62]. Adapted with permission from [62]. © 2020 The Authors. Advanced Intelligent Systems published by Wiley-VCH GmbH.

learning has initiated exchange of ideas between the neuroscience and computing communities, it remains in infancy (Figure 7.12). Brain-inspired neuromorphic computing has the potential to significantly bridge this knowledge gap and pave the pathway for designing efficient AI systems.

References

[1] Bi, G. Q. and Poo, M. M., 1998. Synaptic modifications in cultured hippocampal neurons: Dependence on spike timing, synaptic strength, and postsynaptic cell type. *Journal of Neuroscience*, 18(24), pp. 10464–10472.

[2] Lu, S. and Sengupta, A., 2022, June. Hybrid neuromorphic systems: An algorithm-application-hardware-neuroscience co-design perspective: Invited special session paper. In *2022 IEEE 4th International Conference on Artificial Intelligence Circuits and Systems (AICAS)* (pp. 210–213). IEEE.

[3] Woodin, M. A., Ganguly, K. and Poo, M. M., 2003. Coincident pre-and postsynaptic activity modifies GABAergic synapses by postsynaptic changes in Cl – Transporter activity. *Neuron*, 39(5), pp. 807–820.

[4] Serrano-Gotarredona, T., Masquelier, T., Prodromakis, T., Indiveri, G. and Linares-Barranco, B., 2013. STDP and STDP variations with memristors for spiking neuromorphic learning systems. *Frontiers in Neuroscience*, 7, p. 2.

[5] Diehl, P. U. and Cook, M., 2015. Unsupervised learning of digit recognition using spike-timing-dependent plasticity. *Frontiers in Computational Neuroscience*, 9, p. 99.

[6] Lee, C., Srinivasan, G., Panda, P. and Roy, K., 2018. Deep spiking convolutional neural network trained with unsupervised spike-timing-dependent plasticity. *IEEE Transactions on Cognitive and Developmental Systems*, 11(3), pp. 384–394.

[7] Werbos, P. J., 1990. Backpropagation through time: What it does and how to do it. *Proceedings of the IEEE*, 78(10), pp. 1550–1560.

[8] Lin, J., Lu, S., Bal, M. and Sengupta, A., 2024. Benchmarking spiking neural network learning methods with varying locality. arXiv preprint arXiv:2402.01782.

[9] Bellec, G., Salaj, D., Subramoney, A., Legenstein, R. and Maass, W., 2018. Long short-term memory and learning-to-learn in networks of spiking neurons. *Advances in Neural Information Processing Systems*, 31, pp. 1–11.

[10] Neftci, E. O., Mostafa, H. and Zenke, F., 2019. Surrogate gradient learning in spiking neural networks: Bringing the power of gradient-based optimization to spiking neural networks. *IEEE Signal Processing Magazine*, 36(6), pp. 51–63.

[11] Diehl, P. U., Neil, D., Binas, J., Cook, M., Liu, S. C. and Pfeiffer, M., 2015, July. Fast-classifying, high-accuracy spiking deep networks through weight and threshold balancing. In *2015 International Joint Conference on Neural Networks (IJCNN)* (pp. 1–8). IEEE.

[12] Sengupta, A., Ye, Y., Wang, R., Liu, C. and Roy, K., 2019. Going deeper in spiking neural networks: VGG and residual architectures. *Frontiers in Neuroscience*, 13, p. 95.

[13] Rathi, N., Srinivasan, G., Panda, P. and Roy, K., 2020. Enabling deep spiking neural networks with hybrid conversion and spike timing dependent backpropagation. arXiv preprint arXiv:2005.01807.

[14] Lu, S. and Sengupta, A., 2022. Neuroevolution guided hybrid spiking neural network training. *Frontiers in Neuroscience*, 16, p. 838523.

[15] Lillicrap, T. P., Cownden, D., Tweed, D. B. and Akerman, C. J., 2016. Random synaptic feedback weights support error backpropagation for deep learning. *Nature Communications*, 7(1), p. 13276.

[16] Baldi, P., Sadowski, P. and Lu, Z., 2017. Learning in the machine: The symmetries of the deep learning channel. *Neural Networks*, 95, pp. 110–133.

[17] Lillicrap, T. P., Santoro, A., Marris, L., Akerman, C. J. and Hinton, G., 2020. Backpropagation and the brain. *Nature Reviews Neuroscience*, 21(6), pp. 335–346.

[18] Bellec, G., Scherr, F., Subramoney, A., Hajek, E., Salaj, D., Legenstein, R. and Maass, W., 2020. A solution to the learning dilemma for recurrent networks of spiking neurons. *Nature Communications*, 11(1), p. 3625.

[19] Kaiser, J., Mostafa, H. and Neftci, E., 2020. Synaptic plasticity dynamics for deep continuous local learning (DECOLLE). *Frontiers in Neuroscience*, 14, p. 424.

[20] Gerstner, W., Lehmann, M., Liakoni, V., Corneil, D. and Brea, J., 2018. Eligibility traces and plasticity on behavioral time scales: Experimental support of neoHebbian three-factor learning rules. *Frontiers in Neural Circuits*, 12, p. 53.

[21] Cortes, C., Mohri, M. and Rostamizadeh, A., 2012. Algorithms for learning kernels based on centered alignment. *The Journal of Machine Learning Research*, 13, pp. 795–828.

[22] Li, Y., Kim, Y., Park, H. and Panda, P., 2023. Uncovering the representation of spiking neural networks trained with surrogate gradient. arXiv preprint arXiv:2304.13098.

[23] Maass, W., 2015. To spike or not to spike: That is the question. *Proceedings of the IEEE*, 103(12), pp. 2219–2224.

[24] Ardakani, A., Ardakani, A. and Gross, W. J., 2021, October. Fault-tolerance of binarized and stochastic computing-based neural networks. In *2021 IEEE Workshop on Signal Processing Systems (SiPS)* (pp. 52–57). IEEE.

[25] Roy, K., Sengupta, A. and Shim, Y., 2018. Perspective: Stochastic magnetic devices for cognitive computing. *Journal of Applied Physics*, *123*(21), pp. 210901-1–210901-11.

[26] Roy, D., Chakraborty, I. and Roy, K., 2019, July. Scaling deep spiking neural networks with binary stochastic activations. In *2019 IEEE International Conference on Cognitive Computing (ICCC)* (pp. 50–58). IEEE.

[27] Sengupta, A., Parsa, M., Han, B. and Roy, K., 2016. Probabilistic deep spiking neural systems enabled by magnetic tunnel junction. *IEEE Transactions on Electron Devices*, *63*(7), pp. 2963–2970.

[28] Islam, A. N. M., Saha, A., Jiang, Z., Ni, K. and Sengupta, A., 2023. Hybrid stochastic synapses enabled by scaled ferroelectric field-effect transistors. *Applied Physics Letters*, *122*(12), pp. 123701-1–123701-7.

[29] Bagheri, A., Simeone, O. and Rajendran, B., 2018, April. Training probabilistic spiking neural networks with first-to-spike decoding. In *2018 IEEE International Conference on Acoustics, Speech and Signal Processing (ICASSP)* (pp. 2986–2990). IEEE.

[30] Jiang, Y., Lu, S. and Sengupta, A., 2024. Stochastic spiking neural networks with first-to-spike coding. arXiv preprint arXiv:2404.17719.

[31] Yang, K., Gm, D. P. and Sengupta, A., 2023. Leveraging probabilistic switching in superparamagnets for temporal information encoding in neuromorphic systems. *IEEE Transactions on Computer-Aided Design of Integrated Circuits and Systems*, *42*(10), pp. 3464–3468.

[32] Bengio, Y., Léonard, N. and Courville, A., 2013. Estimating or propagating gradients through stochastic neurons for conditional computation. arXiv preprint arXiv:1308.3432.

[33] McCloskey, M. and Cohen, N.J., 1989. Catastrophic interference in connectionist networks: The sequential learning problem. *Psychology of Learning and Motivation* 24, pp. 109–165.

[34] Kirkpatrick, J., Pascanu, R., Rabinowitz, N., Veness, J., Desjardins, G., Rusu, A. A., Milan, K., Quan, J., Ramalho, T., Grabska-Barwinska, A. and Hassabis, D., 2017. Overcoming catastrophic forgetting in neural networks. *Proceedings of the National Academy of Sciences*, *114*(13), pp. 3521–3526.

[35] Zhang, H. T., Park, T. J., Islam, A. N., Tran, D. S., Manna, S., Wang, Q., Mondal, S., Yu, H., Banik, S., Cheng, S. and Zhou, H., 2022. Reconfigurable perovskite nickelate electronics for artificial intelligence. *Science*, *375*(6580), pp. 533–539.

[36] Ernst, A., Alkass, K., Bernard, S., Salehpour, M., Perl, S., Tisdale, J., Possnert, G., Druid, H. and Frisén, J., 2014. Neurogenesis in the striatum of the adult human brain. *Cell*, *156*(5), pp. 1072–1083.

[37] Arzate, D. M. and Covarrubias, L., 2020. Adult neurogenesis in the context of brain repair and functional relevance. *Stem Cells and Development*, *29*(9), pp. 544–554.

[38] Parisi, G. I., Kemker, R., Part, J. L., Kanan, C. and Wermter, S., 2019. Continual lifelong learning with neural networks: A review. *Neural Networks*, *113*, pp. 54–71.

[39] Stanley, K. O., Clune, J., Lehman, J. and Miikkulainen, R., 2019. Designing neural networks through neuroevolution. *Nature Machine Intelligence*, *1*(1), pp. 24–35.

[40] Marsland, S., Shapiro, J. and Nehmzow, U., 2002. A self-organising network that grows when required. *Neural Networks*, *15*(8–9), pp. 1041–1058.

[41] Bal, M. and Sengupta, A., 2024, May. Equilibrium-based learning dynamics in spiking architectures. In *2024 IEEE International Symposium on Circuits and Systems (ISCAS)* (pp. 1–5). IEEE.

[42] Hinton, G. E., 2002. Training products of experts by minimizing contrastive divergence. *Neural Computation*, *14*(8), pp. 1771–1800.

[43] Berkes, P., Orbán, G., Lengyel, M. and Fiser, J., 2011. Spontaneous cortical activity reveals hallmarks of an optimal internal model of the environment. *Science*, *331*(6013), pp. 83–87.

[44] Scellier, B. and Bengio, Y., 2017. Equilibrium propagation: Bridging the gap between energy-based models and backpropagation. *Frontiers in Computational Neuroscience*, *11*, p. 24.

[45] Xiao, M., Meng, Q., Zhang, Z., Wang, Y. and Lin, Z., 2021. Training feedback spiking neural networks by implicit differentiation on the equilibrium state. *Advances in Neural Information Processing Systems*, *34*, pp. 14516–14528.

[46] Ernoult, M., Grollier, J., Querlioz, D., Bengio, Y. and Scellier, B., 2019. Updates of equilibrium prop match gradients of backprop through time in an RNN with static input. *Advances in Neural Information Processing Systems*, *32*, pp. 1–11.

[47] Pineda, F., 1987. Generalization of back propagation to recurrent and higher order neural networks. In *Neural Information Processing Systems*.

[48] Martin, E., Ernoult, M., Laydevant, J., Li, S., Querlioz, D., Petrisor, T. and Grollier, J., 2021. EqSpike: Spike-driven equilibrium propagation for neuromorphic implementations. *Iscience*, *24*(3), pp. 102222-1–102222-12.

[49] Bal, M. and Sengupta, A., 2022. Sequence learning using equilibrium propagation. arXiv preprint arXiv:2209.09626.

[50] Ramsauer, H., Schäfl, B., Lehner, J., Seidl, P., Widrich, M., Adler, T., Gruber, L., Holzleitner, M., Pavlović, M., Sandve, G. K. and Greiff, V., 2020. Hopfield networks is all you need. arXiv preprint arXiv:2008.02217.

[51] Krotov, D. and Hopfield, J., 2018. Dense associative memory is robust to adversarial inputs. *Neural Computation*, *30*(12), pp. 3151–3167.

[52] Lin, J., Bal, M. and Sengupta, A., 2024. Scaling SNNs trained using equilibrium propagation to convolutional architectures. arXiv preprint arXiv:2405.02546.

[53] Bai, S., Kolter, J. Z. and Koltun, V., 2019. Deep equilibrium models. *Advances in Neural Information Processing Systems*, *32*, pp. 1–12.

[54] Bal, M. and Sengupta, A., 2024, March. SpikingBERT: Distilling BERT to train spiking language models using implicit differentiation. In *Proceedings of the AAAI Conference on Artificial Intelligence* (Vol. 38, No. 10, pp. 10998–11006).

[55] Bal, M., Jiang, Y. and Sengupta, A., 2024. Exploring extreme quantization in spiking language models. arXiv preprint arXiv:2405.02543.

[56] Lu, S. and Sengupta, A., 2022, June. Hybrid neuromorphic systems: An algorithm-application-hardware-neuroscience co-design perspective: Invited special session paper. In *2022 IEEE 4th International Conference on Artificial Intelligence Circuits and Systems (AICAS)* (pp. 210–213). IEEE.

[57] Lee, C., Panda, P., Srinivasan, G. and Roy, K., 2018. Training deep spiking convolutional neural networks with STDP-based unsupervised pre-training followed by supervised fine-tuning. *Frontiers in Neuroscience*, *12*, p. 435.

[58] Feoktistov, V., 2006. *Differential evolution*. New York: Springer US.

[59] Lu, S. and Sengupta, A., 2022. Neuroevolution guided hybrid spiking neural network training. *Frontiers in Neuroscience*, *16*, p. 838523.

[60] Singh, S., Sarma, A., Jao, N., Pattnaik, A., Lu, S., Yang, K., Sengupta, A., Narayanan, V. and Das, C. R., 2020, May. Nebula: A neuromorphic spin-based ultra-low power architecture for SNNs and ANNs. In *2020 ACM/IEEE 47th Annual International Symposium on Computer Architecture (ISCA)* (pp. 363–376). IEEE.

[61] Pei, J., Deng, L., Song, S., Zhao, M., Zhang, Y., Wu, S., Wang, G., Zou, Z., Wu, Z., He, W. and Chen, F., 2019. Towards artificial general intelligence with hybrid Tianjic chip architecture. *Nature*, *572*(7767), pp. 106–111.

[62] Aimone, J. B., 2021. A roadmap for reaching the potential of brain-derived computing. *Advanced Intelligent Systems*, *3*(1), p. 2000191.

[63] Han, Z., Islam, A. N. and Sengupta, A., 2023, June. Astromorphic self-repair of neuromorphic hardware systems. In *Proceedings of the AAAI Conference on Artificial Intelligence* (Vol. 37, No. 6, pp. 7821–7829).

[64] London, M. and Häusser, M., 2005. Dendritic computation. *Annual Review of Neuroscience*, *28*(1), pp. 503–532.

[65] Plagge, M., Cardwell, S. G. and Chance, F. S., 2024, April. Expressive dendrites in spiking networks. In *2024 Neuro Inspired Computational Elements Conference (NICE)* (pp. 1–8). IEEE.

[66] Goan, E. and Fookes, C., 2020. Bayesian neural networks: An introduction and survey. *Case Studies in Applied Bayesian Data Science: CIRM Jean-Morlet Chair, Fall 2018*, pp. 45–87.

8

Lifelong Learning with AI Algorithms and Hardware

8.1 Emulating Complex Neural Functions beyond Plasticity

Plasticity of synapses is often considered the primary mechanism for learning and is widely used in training artificial neural networks. At the same time, it is well known that backpropagation which is utilized for error mitigation in ANNs is not biologically plausible due to the large number of connections between neuronal units in the brain. Several mechanisms for learning have been identified and put forward that go beyond the Hebbian concept of time-dependent activity of presynaptic and postsynaptic neurons. For instance, in recent years, computation within neurons while considering the complexity of their dendritic morphology and internal structure is emerging as an important area of research. In fact, recent studies have suggested that individual layer 5 cortical pyramidal neuron cell can be as sophisticated as a five-to-eight-layer deep neural network based on the input–output mapping [1]. Clearly, the implication is that simple, linear summation of synaptic inputs by a neuron to produce an output (e.g., action potential) is not entirely accurate. The complexity of the computation within the neuron considering the role of dendritic trees, time dependence of signal input, and spatial locations of synaptic connections, should all be considered both for interpreting the experimental measurements from neurons in animal models and design of artificial neurons for neuromorphic hardware. In fact, the idea of dendrocentric intelligence has also been proposed as a possible approach to reduce energy consumption in ANNs for training [2]. Figure 8.1 illustrates the possible contributions from dendrites and synaptic inputs located at different regions in a neuron to the overall generation of an action potential [3].

The shape of the dendritic plateau potential is much flatter than what is normally seen in a neuronal action potential, this indicates the possibility to hold longer term memory for subsequent activation of a neuron. Synaptic inputs from different junctions could add in a nonlinear manner depending on their spatial location and

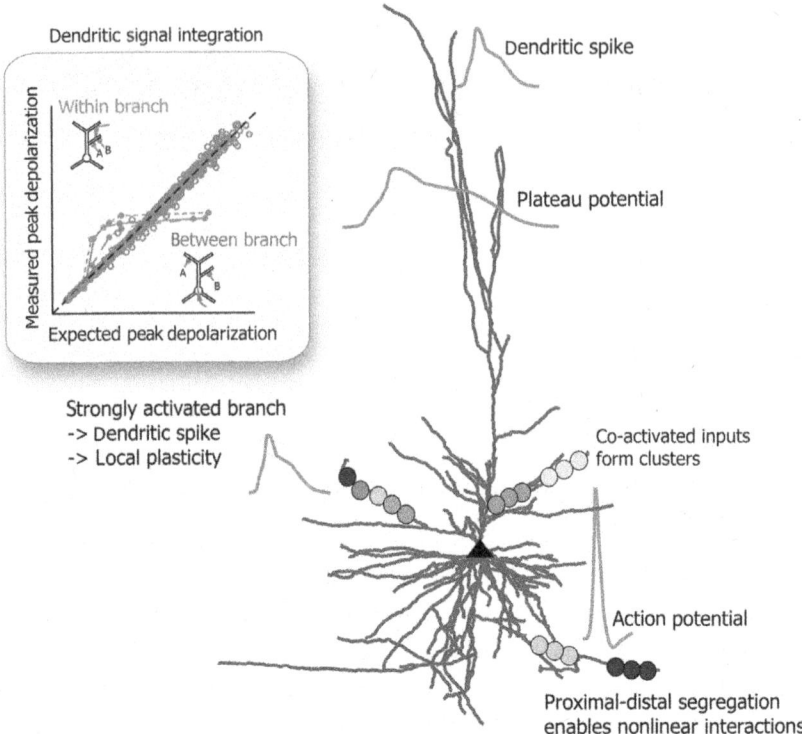

Figure 8.1 Spikes can originate at dendrites from different locations in a neuronal cell. The width of the dendritic plateau potentials can be longer than typical action potentials. Signal integration can be quite nonlinear, for instance, signal integration within an activated branch can be sigmoidal while summation across branches can be linear. The circles in the dendrites represent synapses that are formed in clusters. Synapses that are activated at different distances from the cell body also can sum differently. The local structure and spatial distribution of the synaptic inputs in the dendrites therefore results in distinct summation rules and plasticity within a single neuron cell [3]. Adapted with permission from [3]. © 2021 IBRO. Published by Elsevier Ltd. All rights reserved.

time interval between activation of the synapses from adjacent neurons. The integration of signals can deviate significantly from linear addition depending on the spatial proximity in the dendritic branches. Such examples of signal summation within a neuron indicates a complexity that far exceeds the trivial assumption of the neuron as a simple adder unit that fires upon reaching a threshold value of the membrane potential. Local generation of dendritic spikes opens the possibility to implement spatially limited learning rules that do not require an action potential to be generated at the cell body. Hence, synapses connected to different regions in a neuron can be modified in a selective manner. In other words, a whole gamut of

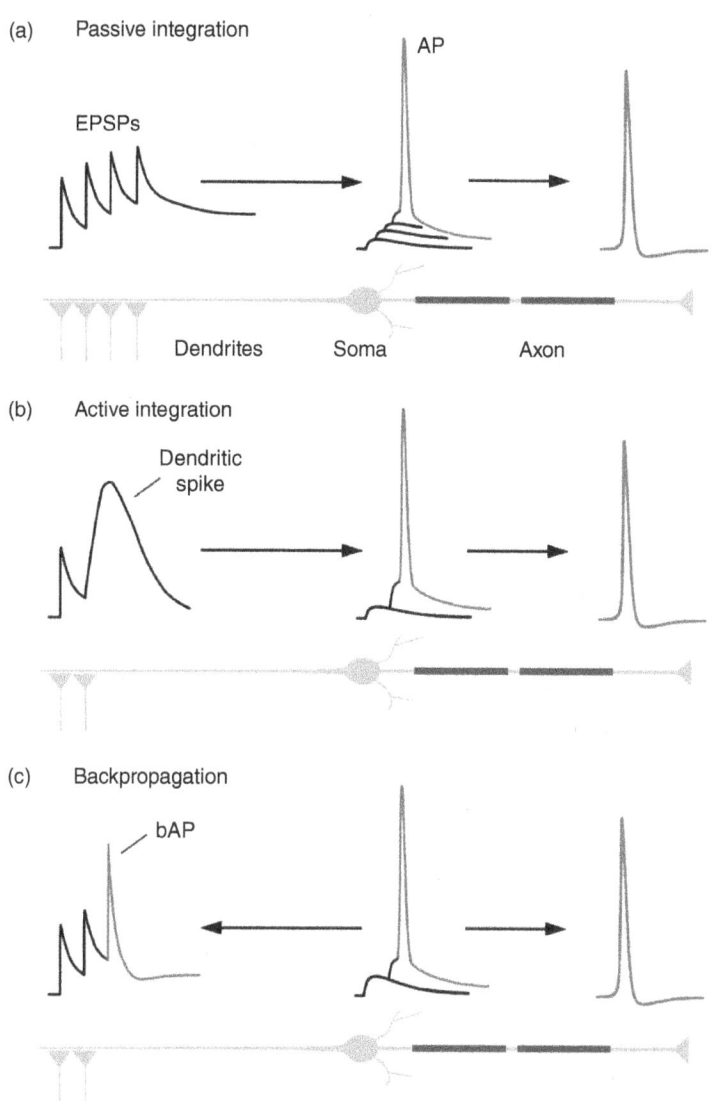

Figure 8.2 (a) Passive integration, (b) active integration, and (c) backpropagation of action potentials in dendrites. A range of computational capabilities have been noted in dendrites including logical operations, low-pass filtering, amplification of signals, and coincidence detection. Dendrites can amplify or reduce the interaction between synapses by adding signals in a sublinear or linear manner depending on the spatial location of the excitatory inputs [4]. Adapted with permission from [4]. Copyright © 2015, Springer Nature America, Inc.

novel learning algorithms can be implemented by understanding the morphological diversity of neurons. A snapshot of the diversity of dendritic signal processing is summarized in Figure 8.2 [4].

These observations in neuroscience have significant implications in neuromorphic computing. From the algorithm's perspective, the neuron cannot simply be considered as a linear sum of synaptic inputs for thresholding. Mathematically tractable temporal summation rules and local plasticity models should be implemented considering the specific morphology of the neuron [5]. Due to the broad plateau potentials, history effects and memory of recent activity could be also factored in. Coincidence detection can be realized within a neuron due to backpropagation of action potential that can synchronize with a distal synaptic excitation. From the hardware perspective, clearly there needs to be an effort to emulate the dendritic functions beyond just considering neurons as linear summation units. Calcium-mediated dendritic action potentials (dCaAPs) have inspired construction of artificial neuronal devices based on insulator–metal transition phenomenon that shows negative differential resistance (NDR). The non-monotonous current–voltage relationship enables realization of the output voltage to reach a peak value at an intermediate value of current. This characteristic allows implementation of XOR logic like what has been noted in dendrites in cortical neurons [6]. Integration of such devices with nonvolatile memory can help realize in-memory computing which is a promising direction for energy-efficient non-von Neumann processors. In the field of mixed-signal neuromorphic computing hardware, dendritic features of the neuronal subunits are beginning to be implemented in large-scale systems [7]. Dendritic circuits have also been fabricated with resistive memory technologies integrated onto CMOS platforms to demonstrate spatiotemporal feature detection [8].

8.2 Adaptation, Learning, and Evolutionary Dynamics

Neural circuits have the unique ability to adapt to recurring stimuli and learn through the lifespan. Implementing subsets of these features into artificial neural networks is often accomplished at the synapse level by tuning the weights (referred to as synaptic plasticity). However, this simplistic assignment of learning features restricted to just synapses does not capture the full spectrum.

Figure 8.3 shows both non-synaptic and synaptic changes that can mediate learning and memory [9]. While the change in synaptic weight due to activity at the pre- and post-neuron is reminiscent of Hebbian plasticity, other mechanisms can also be at play. Non-synaptic effects include modification of neuronal excitability by several means including modifying the resting membrane potential, increasing the resistance, and modification of the ionic currents. The changes to the neuron properties can also serve as a memory trace for the subsequent interactions with synapses or their activation; hence learning properties can be shared between neurons and synapses in addition to maintaining homeostasis. Again, it is important to note that the learning properties can be site-specific and occur at dendritic branches.

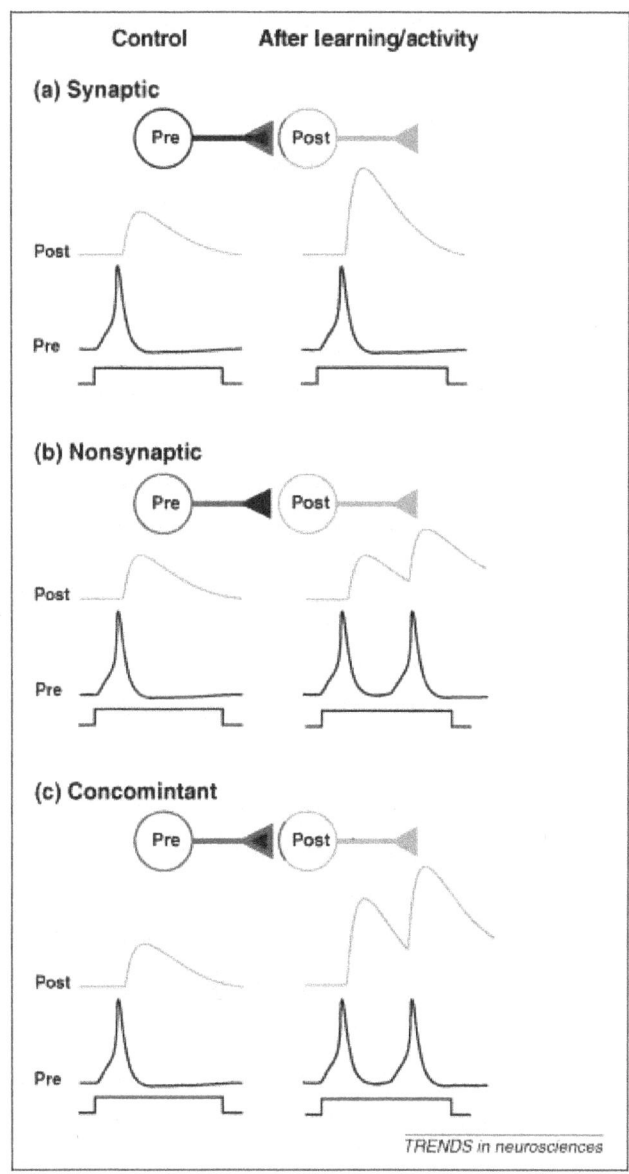

Figure 8.3 Synaptic and non-synaptic changes that can lead to learning. (a) Shows changes in synaptic weight that modifies the interaction between a presynaptic and postsynaptic cell, (b) shows changes in excitability of neuronal cells by generating greater number of action potentials, and (c) shows combination of both synaptic and non-synaptic effects. Increased activity of the neurons in turn invokes greater changes to the synapse [9]. Adapted with permission from [9]. © 2009 Elsevier Ltd. All rights reserved.

In other words, learning can be spatially localized mediated by changes in neuron properties such as number and conductance of ion channels and/or in combination with changes to the synapses via neurotransmitters or generation of proteins [9].

So, how can these features be adapted into neuromorphic algorithms and hardware? One would need to look beyond the point neuron approach and begin to integrate features of learning into the various components that go into making the neural network fabric. In most approaches to date, memory devices are used for implementing learning and plasticity, such as potentiation–depression characteristics while the neuron is simply used for thresholding. As discussed in Section 8.1, incorporating finer features into the algorithms and artificial neuron hardware such as dendritic spines, spatial inhomogeneity of electronic properties, or imposing persistent gradients across the physical device might be required in future. As an example, introducing thermal gradients across a neuron device could result in local generation of action potentials at specific sites that reach a threshold temperature.

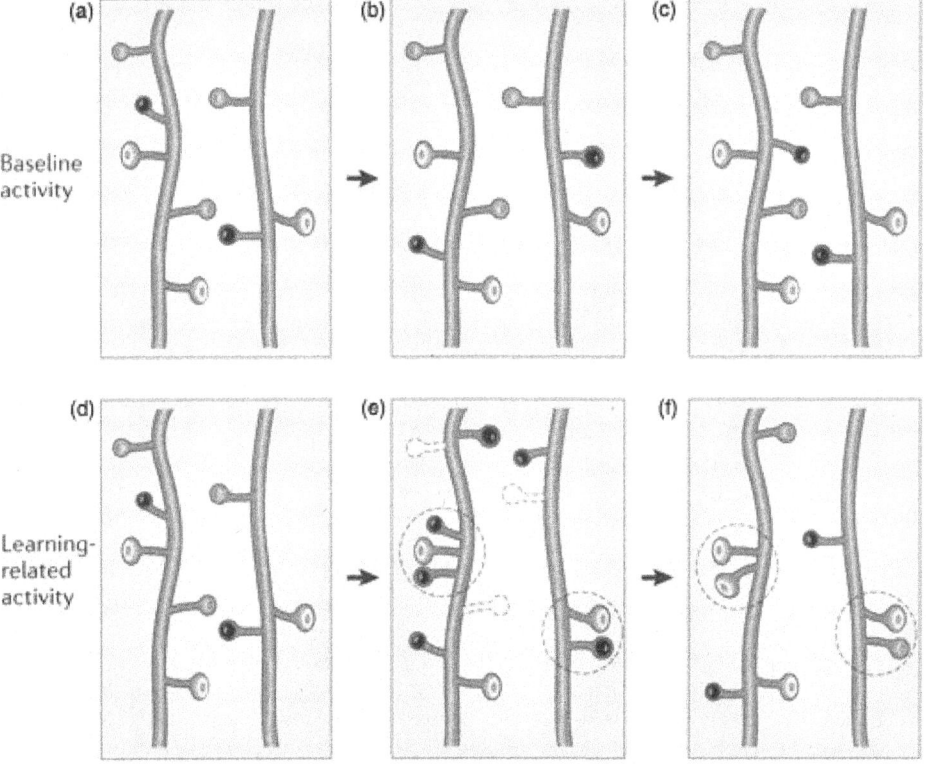

Nature reviews | Neuroscience

Figure 8.4 (a–c) Baseline activity of dendritic spine structural changes. Small dark spines are transient in this case. (d–f) Under learning conditions, new spines are formed while preexisting ones can disappear. Interestingly, new spines are formed in clusters and can become stabilized indicating a mechanism for long-term memory formation [10]. Adapted with permission from [10]. Copyright © 2012, Springer Nature Limited.

This would involve creating more complex device architectures beyond two terminals to emulate dendritic junctions and multiple locations for synaptic inputs and feedback. Alternatively, such devices must be colocated on chip with predefined heat transfer characteristics. In other words, there is a trade-off between device structure simplicity, circuit architecture, and implementing finer aspects of biological learning mechanisms particularly plasticity.

Memory traces can be implemented in other ways as well. Structural plasticity of neural components is an important mechanism in this regard, namely, the dynamic nature of connectivity of synapses [10]. Changes due to experience-dependent plasticity in a growing brain can occur besides learning or recovery post-injury in a mature adult brain [11]. Synapse formation and elimination, and reorganization of interconnections with neurons affect neural circuit function and therefore important for learning and decision-making. For instance, size and shape of dendritic spines can change via induction of synaptic plasticity. The structural changes can in turn help stabilize a synapse during training such as has been noted in motor skill learning and fear conditioning experiments. Figure 8.4 provides an illustration of structural changes that happen in synaptic networks due to behavioral learning [10].

The structural plasticity including growth of new spines could occur with delays of up to tens of hours indicating mechanisms of memory consolidation and long-term effects. These observations pose grand challenges to design of neuromorphic hardware for obvious reasons. In integrated circuit technologies, we are used to precisely fabricating devices and their interconnections and operate them until end of life. Any changes to the connections typically result in malfunction or failure of a device or chip. In biology on the other hand, neural circuits are entirely dynamic and change through the lifespan of an organism. So, how can we reconcile this stark contrast in neuromorphic computing? On the machine learning side, continual learning and avoiding catastrophic forgetting is an active area of research [12]. Implementing structural plasticity requires a fundamental rethink in hardware design, and we briefly discuss this in Section 8.3.

8.3 Bioinspired versus Bio-realistic Functions in AI Algorithms and Hardware

As discussed throughout this book, one primary goal of neuromorphic computing is to design hardware and algorithms that are grounded in biological neural circuit behavior. Clearly this can be interpreted at various levels. For instance, nature-inspired algorithms are widely used in computer science. Swarm intelligence is among the most prominent in this category, wherein collective behavior of insects, animals, or birds inspire algorithm development. Ant colonies are perhaps among the most studied in this context. Mathematical rules that can explain collective response of a colony of animals or a flock of birds can be abstracted for designing an algorithm to solve some specific problem. It is possible to extend this analogy to semiconductor materials and devices as well. Just as each ant in a colony represents a distinct entity that is part of a larger network, it is possible to argue that electrons in crystal lattices represent components and collectively their interactions result in a nontrivial output. In this scenario, neuromorphic hardware is not utilized directly in building a circuit but their behavior under external stimulus can be utilized to design a learning algorithm. One example is implementation of adaptive synaptic plasticity in neural networks based on habituation behavior demonstrated by a quantum electronic device [13]. Learning characteristics observed in a solid-state system were abstracted into a set of equations that was utilized in neural network plasticity algorithms to demonstrate gradual forgetting. Based on the specific device technology such as insulator–metal transition or magnetic ordering, different equations can fit the electrical behavior and can be tested for optimal performance in networks. This direction of research, namely investigating hardware equivalents of *Aplysia* or other model systems such as *C. elegans* in neuroscience offers a novel venue to examine emerging semiconductors for artificial intelligence in a broadly defined manner.

Even when semiconductor devices are used to build neuromorphic circuits and connected with peripheral CMOS circuitry to demonstrate prototype processors, the extent of biological relevance can vary. In the biological neural networks, all currents are ionic as we have discussed already. The unequal amounts of ionic species or their diffusivity across a cell membrane establish a baseline voltage (resting potential) which is perturbed by ionic current flow. Similarly, for short-term versus long-term plasticity at a synapse, proteins could be synthesized at the junctions. A variety of neurotransmitters are constantly being generated and consumed in the brain matter. Neuromorphic hardware do not resemble any of their biological counterparts in this regard. Most hardware rely on electron flow including the ones wherein ionic species are migrating. As we have discussed already, neuron function is captured by a thresholding signal generated by a device while a synapse refers to a memory state that can be tuned. The bio-inspiration is from emulating the observed macroscopic signals in biological networks such as an action potential train and/or time-dependent EPSP variation at a synapse junction. Similarly, the rules that are implemented in the networks to modify the weights (e.g., STDP) have some basis in neuroscience. In fact, in early demonstrations of silicon CMOS-based neural networks, special circuitry had to be utilized to slow down the operational speeds to match biological dynamics.

Bio-realistic devices take this one step further and attempt to emulate the finer features of biological networks. For instance, artificial neurons that can demonstrate self-oscillations mimic the periodic generation of action potentials in neuronal cells. However, biological neurons also have several additional characteristics, including tonic spiking, bursting, subthreshold oscillations, and stochastic spiking. To emulate such fine features, it is necessary to look for devices that host complex intrinsic physics. For instance, devices that require just one state variable (first order) are in-sufficient to replicate the multitude of characteristics demonstrated by biological neurons. It is essential to move up to second-, third-, or fourth-order devices for these purposes. Additional state variables, such as concentration-dependent mobility of charge carriers, nonlinear temperature dependence of carrier density, and dopant distribution in space-dependent global device conductance are some examples of complexity that are needed. Using IMT materials such as NbO_2 and VO_2, researchers have shown bio-realistic neuronal functions and in some cases, it requires more than one device element. In a recent study, using VO_2 switches, 23 types of neuronal behaviors including bursting, spike frequency adaptation, threshold variability, and refractory period were demonstrated [14]. Similar classes of quantum materials have also been used to demonstrate devices that operate in chaotic modes. Increasing complexity at the device level can ensure realization of complex functions noted in biology. Without doubt, CMOS transistors are complex but are simple from the perspective of temperature dependence within the operating boundary conditions (i.e., mostly insensitive to small temperature perturbations), nearly perfect from a silicon wafer structural point of view and fabricated to switch with negligible variability and ultrahigh probability. What we are discussing in neuromorphic hardware to achieve bio-realism at low footprint on the other hand is a significant departure from the existing paradigm of semiconductor device manufacturing. Strong temperature dependence that results in nonlinear signal output, extreme concentrations of impurities that are subject to strong field-driven drift, probabilistic switching are all routinely implemented in emerging neuromorphic hardware and prized! In fact, recent studies on probabilistic computing hardware known as *p*-bits show that stochastic nature of state switching and coupling between junctions can significantly boost the computing efficiency and power savings [15, 16]. It is going to take some time for the traditional semiconductor industry technologists to adapt to emerging hardware for neuromorphic computing using non-silicon devices. Indeed, it is not surprising that at the time of writing this book, most industrial neuromorphic chips are still manufactured using silicon CMOS technologies.

8.4 Future Directions in Biomimetic Computation

Enormous progress has been made in design and experimental fabrication of neuromorphic hardware over the past few decades. Starting from early demonstrations of subthreshold silicon transistor circuits to emulate neuronal function, actively tunable materials are now able to mimic the 1907 Lapicque model of the neuron in just two-terminal device configurations. Numerous materials classes can now realize basic synapse functions such as nonvolatile memory with adaptive and multiple states while consuming low power for state setting. At the same time, much research is needed to advance the frontiers of this field and can be broken down into elementary categories. Starting at the material level, memory is introduced through defects primarily in most of the experimental approaches to date. The defect concentration is such that we are in the non-dilute regime in many of the scenarios and moreover spatially distributed in a highly nonuniform manner such as in filamentary switches. In ferroelectric systems, extended defects such as domain boundaries are involved in addition to point defects such as oxygen vacancies that can migrate under electric fields. Hence, a deep understanding of defect migration, spatial characterization in operando devices, and microscopic description of their effects on charge transport are essential. The materials are further compositionally and structurally complex (e.g., polycrystalline) in most cases, especially when they are integrated onto backend of silicon circuits. The active layers are often heterogenous semiconductor–metal phase mixtures; hence manufacturing issues such as reproducible and scalable fabrication need to be addressed. It is important to point out the active materials being considered for neuromorphic computing are quite different from the pristine ultrapure, ultralow defect single crystal Si or GaAs wafers one comes across in basic semiconductor courses, physics textbooks, and today's semiconductor industry. The only exception is the continued use of silicon CMOS circuits repurposed for neuromorphic computing, but as we have discussed already, they are far from being bio-realistic and not energy-efficient.

At the device level, two-terminal devices have taken over a significant portion of research in neuromorphic hardware. In part, due to their simpler design compared to three-terminal transistor type structures and the ability to fabricate them with limited lithography capabilities, such devices are being proposed for a variety of neuromorphic technologies. In almost all cases, the channel region is an electronically inhomogeneous medium. This implies a complex distribution of electric field pattern inside the device and its variation during operation, reset, and cycling. On the one hand, such complex features enable higher order device function as outlined earlier, but also come with challenges such as stochastic operation and unknown size effects. If the variability cannot be minimized to acceptable levels, certain neuromorphic architectures will simply not be possible to realize such as large-scale synchronized oscillators. On the other hand, probabilistic computing could indeed benefit from stochastic switches. There is a strong analogy in biology for operating with inherent variability. Hence, a careful analysis of use case scenarios must be determined to identify suitable applications for emerging device technologies.

8.4 Future Directions in Biomimetic Computation

At the algorithm and architecture levels lies some of the greatest unknowns currently. In neuroscience, it is well known that both the number of neurons and synapses, number density as well as their topology are important for cognitive capabilities. In most experimental works to date, demonstrated circuits are elementary compared to the density of interconnections found in mammalian brains. Even commercial efforts to realize millions of neurons on chips are a far cry from tens of billions and trillions of synaptic connections noted in the human brain. Despite this limitation, it is still possible to of course investigate well-defined tasks such as image recognition, classification of datasets, and learning novel stimuli with small-scale neuromorphic hardware. It is further possible to benchmark their

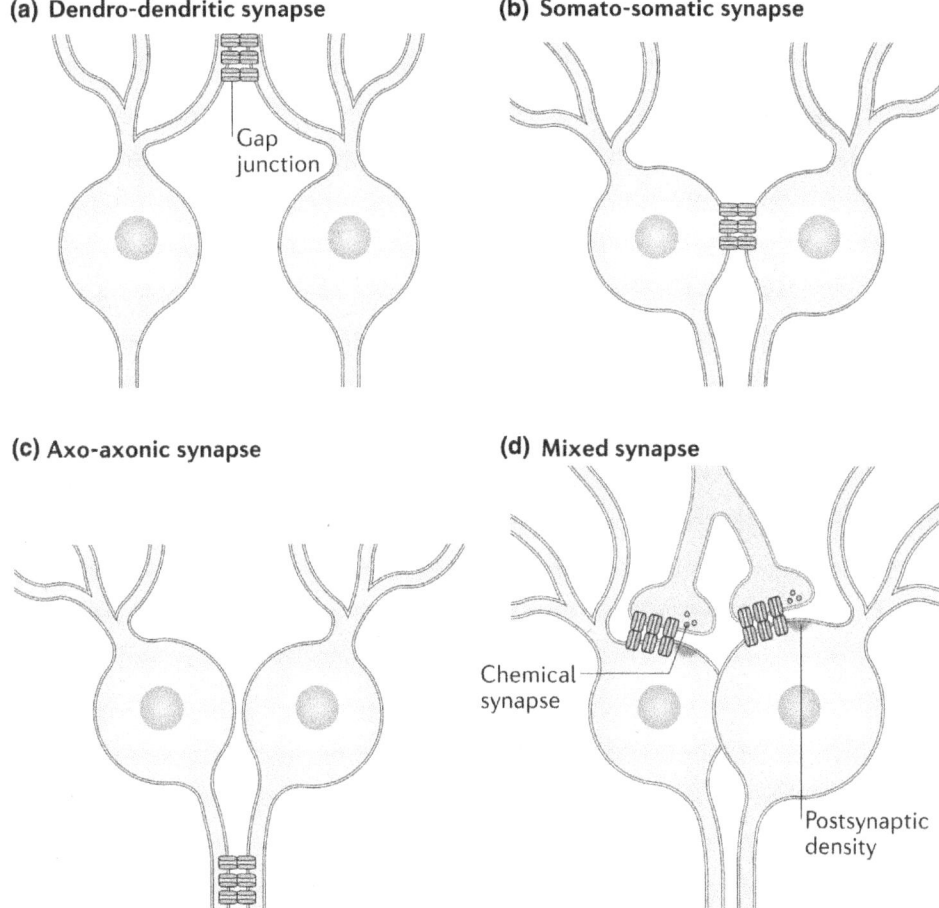

Figure 8.5 (a–d) Electrical synapses (gap junctions) between neurons. The location of the gap junction can vary in the neuronal stem, such as at dendrites, soma, axons, or can be of mixed nature with chemical synapses colocated. Each of these synaptic connections have unique function in the neural circuit [17]. Adapted with permission from [17]. Copyright © 2019, Springer Nature Limited.

performance in terms of energy consumption against software or digital technologies. However, we do not yet understand the limits of intelligence possible with neuromorphic algorithms and hardware and if we can indeed reach the energy efficiency of the human brain for comparable problem solving. Pursuing this goal will likely be a frontier area of research in physical sciences and engineering over the course of next several decades.

Finally, we note the role of understanding neuroscience literature in guiding neuromorphic computing. Neuroscience, like any other, is an evolving field. Recent discoveries suggest the remarkable diversity in neuronal and synaptic function, their complex interrelationships, and novel forms of functional and structural plasticity that can occur at various timescales. Figure 8.5 shows the types of electrical synaptic wiring between neurons considering their location with respect to the cell body [17]. Synaptic connections between dendrites are important in the generation of gamma frequencies and can facilitate lateral excitation despite their distance away from the soma. Synapses between soma have stronger effects on cellular activity, while synapses between axons can contribute to synchronous activity such as fast ripples.

Finally, there can also be mixed chemical–electrical synapses that can serve communication in both directions between presynaptic and postsynaptic cells and improve flexibility in information transfer modes. Hence, there is incredible complexity even in just individual connections between two neurons with distinct effects on signal transfer and circuit-level activity. To truly advance neuromorphic computing, engineers, computer scientists, and physicists need to strengthen the interaction with experimental and computational neuroscientists. Note that several problems facing neuromorphic computing, such as the best means to emulate the biological phenomena without sacrificing rigor, are also apparent in neuroscience. Mathematical models of neuronal circuits, experimental methods to probe transient neuronal dynamics, and interpretation of plasticity measurements in brain slices versus *in vivo* in live animals all require approximations and assumptions to varying degrees. In this regard, it will be beneficial for neuromorphic computing researchers to discuss firsthand with neuroscientists on both bio-realistic and mathematically tractable approaches to emulate neural information processing.

References

[1] Beniaguev, D., Segev, I. and London, M., 2021. Single cortical neurons as deep artificial neural networks. *Neuron, 109*(17), pp. 2727–2739.

[2] Boahen, K., 2022. Dendrocentric learning for synthetic intelligence. *Nature, 612*(7938), pp. 43–50.

[3] Acharya, J., Basu, A., Legenstein, R., Limbacher, T., Poirazi, P. and Wu, X., 2022. Dendritic computing: Branching deeper into machine learning. *Neuroscience, 489*, pp. 275–289.

[4] Stuart, G. J. and Spruston, N., 2015. Dendritic integration: 60 years of progress. *Nature Neuroscience*, *18*(12), pp. 1713–1721.

[5] Poirazi, P. and Papoutsi, A., 2020. Illuminating dendritic function with computational models. *Nature Reviews Neuroscience*, *21*(6), pp. 303–321.

[6] Yang, Z., Yue, W., Liu, C., Tao, Y., Tiw, P.J., Yan, L., Yang, Y., Zhang, T., Dang, B., Liu, K. and He, X., 2024. Fully hardware memristive neuromorphic computing enabled by the integration of trainable dendritic neurons and high-density RRAM chip. *Advanced Functional Materials*, *34*, p. 2405618.

[7] Kaiser, J., Billaudelle, S., Müller, E., Tetzlaff, C., Schemmel, J. and Schmitt, S., 2022. Emulating dendritic computing paradigms on analog neuromorphic hardware. *Neuroscience*, *489*, pp. 290–300.

[8] D'Agostino, S., Moro, F., Torchet, T., Demirağ, Y., Grenouillet, L., Castellani, N., Indiveri, G., Vianello, E. and Payvand, M., 2024. DenRAM: Neuromorphic dendritic architecture with RRAM for efficient temporal processing with delays. *Nature Communications*, *15*(1), p. 3446.

[9] Mozzachiodi, R. and Byrne, J. H., 2010. More than synaptic plasticity: Role of nonsynaptic plasticity in learning and memory. *Trends in Neurosciences*, *33*(1), pp. 17–26.

[10] Caroni, P., Donato, F. and Muller, D., 2012. Structural plasticity upon learning: Regulation and functions. *Nature Reviews Neuroscience*, *13*(7), pp. 478–490.

[11] Butz, M., Wörgötter, F. and Van Ooyen, A., 2009. Activity-dependent structural plasticity. *Brain Research Reviews*, *60*(2), pp. 287–305.

[12] Parisi, G. I., Kemker, R., Part, J. L., Kanan, C. and Wermter, S., 2019. Continual lifelong learning with neural networks: A review. *Neural Networks*, *113*, pp. 54–71.

[13] Panda, P., Allred, J. M., Ramanathan, S. and Roy, K., 2017. ASP: Learning to forget with adaptive synaptic plasticity in spiking neural networks. *IEEE Journal on Emerging and Selected Topics in Circuits and Systems*, *8*(1), pp. 51–64.

[14] Yi, W., Tsang, K. K., Lam, S. K., Bai, X., Crowell, J. A. and Flores, E. A., 2018. Biological plausibility and stochasticity in scalable VO_2 active memristor neurons. *Nature Communications*, *9*(1), p. 4661.

[15] Park, T. J., Selcuk, K., Zhang, H. T., Manna, S., Batra, R., Wang, Q., Yu, H., Aadit, N. A., Sankaranarayanan, S. K., Zhou, H. and Camsari, K. Y., 2022. Efficient probabilistic computing with stochastic perovskite nickelates. *Nano Letters*, *22*(21), pp. 8654–8661.

[16] Camsari, K. Y., Sutton, B. M. and Datta, S., 2019. P-bits for probabilistic spin logic. *Applied Physics Reviews*, *6*(1), pp. 011305-1–011305-12.

[17] Alcamí, P. and Pereda, A. E., 2019. Beyond plasticity: The dynamic impact of electrical synapses on neural circuits. *Nature Reviews Neuroscience*, *20*(5), pp. 253–271.

9
Practice Problems

1. Consider a voltage source (V) applied to a resistor–capacitor (R–C) circuit. Derive an expression for the time constant of the charging response of the capacitor.
2. Let us examine a simplified version of the Hodgkin–Huxley neuron action potential model. We consider only two conductances, namely, due to Na^+ and K^+ ion currents. (Hodgkin et al. experimentally found that the chloride ion currents can be neglected.) For this scenario, derive an expression for the resting potential (V_{rest}) of a neuron in terms of Na^+ and K^+ conductances G_{Na} and G_K and reversal potentials V_{Na} and V_K, respectively.
3. Consider the case of a spherical neuron cell of diameter p. If we are injecting current $I(t)$ into this neuron, where t represents time, derive an expression for the membrane potential as a function of time. Assume the resting potential of the neuron is V_{rest}. What is the voltage dynamics of the cell for constant current of amplitude I_0? State all variables utilized to derive the general membrane equation.
4. Consider a synapse connected to a neuron. The synapse can be treated as having a slowly varying synaptic input along with an internal battery with a characteristic reversal potential E_{syn}. Derive an expression for the saturation membrane potential.
5. The cable equation is used to derive signal transmission in neurons. Explain how the resistance of the neuronal membrane influences the ionic current flow using the space constant. What are the implications in design of artificial neurons using this analogy?
6. Derive the gradient descent equations noted in Section 2.4 of Chapter 2 for ReLU nonlinearity.
7. Train a 3-layered fully connected neural network on the MNIST dataset with 100 neurons in each hidden layer and compare the accuracies for sigmoid, tanh, and ReLU transfer functions.
8. Train VGG-16 and ResNet-34 architectures on the CIFAR-10 and ImageNet datasets and report the top-1 and top-5 error rates. Implement batch normalization layers to improve the accuracy of the networks.

9. Derive the backpropagation equations mentioned in Eqs. (2.7)–(2.12) of Chapter 2.
10. Derive weight update equations for a fully connected neural network when the weights are initialized to zero.
11. A memristor device comprises an active region which is a dielectric containing numerous metal nanoparticle aggregates that are formed between the two electrodes. Assume electron tunneling across metal nanoparticles, each uniformly separated by gap d, is the primary mechanism for current flow. Derive an expression for the total resistance. In this simple model, what microstructural feature controls the resistance predominantly? Are there any fabrication processes that can be useful to reduce the randomness in the distribution of the nanoparticles?
12. Oxides used as resistive switching synapses (e.g., $Ni_{1-\delta}O$) may contain cation vacancies. Assume we only have neutral cation vacancies present. For the defect equilibria of oxygen incorporating into the oxide from gas phase, derive an expression relating the cation non-stoichiometry (δ) to the oxygen activity (a_{O2}). Will this relationship change if the neutral defect ionizes to create a singly positively charged cation vacancy? Explain your answer.
13. Amorphous to crystalline structural transformation is widely studied in solid systems to design memory technologies. If the bulk free energy gain due to phase change is proportional to the undercooling (i.e., difference between actual temperature T and the transformation temperature T_t), derive an expression relating the nucleation rate of crystals to the temperature T. State all variables.
14. Narrow gap insulators may display Poole–Frenkel (PF) conduction mechanism when subject to strong electric fields and high temperatures. Trapped electrons get excited into the conduction band resulting in a finite current flowing through the insulator and can even lead to sharp conductance transitions by collapsing the band gap at critical fields. Atomic force microscopy can be utilized to measure current density locally as a function of voltage using a sharp metal tip scanned across a thin film surface while the substrate could serve as a counter electrode.
    a. If the film thickness (t) and dielectric constant (ε) are known, prove that it is possible to estimate the local temperature using the PF equation for current density.
    b. For a material with bandgap E_g, transition temperature of T_c, estimate the criteria under which it is possible to rule out Schottky emission as a competing mechanism.
    c. Recalculate solution to (b) if Fermi level pinning is present. State any assumptions.

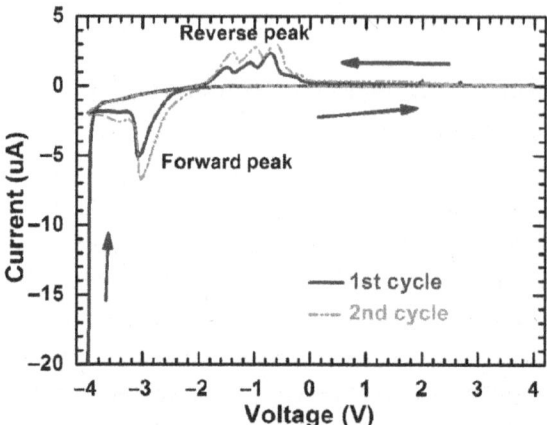

Figure 9.1 Cyclic voltammetry plot for VO$_2$ film interfaced with an ionic liquid. Adapted with permission from [1]. 2012 © AIP Publishing.

15. Ionic liquid or gel interfaces can be used to emulate synaptic properties by control of charge transfer across a solid–liquid interface in electric double layer transistors. Figure 9.1 shows voltage sweep curves for two representative cycles taken from a representative ionic liquid–semiconductor interface. From an analysis of this experimental data, determine whether the forward and reverse polarity reactions are reversible. What does this tell you about the chemical nature of the interface?

16. SrTiO$_3$ is an important material that is both exploited as a substrate for growth of various oxide semiconductors as well as a material that is being explored as a candidate interface-type resistive switching memory material. Following growth of thin film SrTiO$_3$, a user has accidentally left the sample under ultra-high vacuum at elevated temperatures overnight. The next day, upon removing the sample from the growth chamber, the user suspects the quenched sample might be significantly oxygen-deficient leading to a conducting material instead of the expected insulating ground state. Write a possible defect reaction for this scenario, one possible charge neutrality condition and the resulting dependence of carrier density on the oxygen partial pressure if this sample is heated in similar environmental conditions.

17. Next, we consider the case of SrTiO$_3$ film that contains oxygen vacancies due to *extrinsic* factors such as impurities from the precursor target. Assume the user has left the oxygen flow valve ON by mistake for several hours at elevated temperatures. What would be the expected relation between carrier density and oxygen partial pressure if this sample is heated in similar environmental conditions?

18. Ionic gel gating of nearly degenerate semiconductors are widely studied for strong electric field gating, ion intercalation and doping in electrolytic transistors as well as in emerging computing technologies such as artificial synapses and reservoir computing. For the case of a gel gate in contact with one such semiconductor and subject to strong electric fields, electrons and ions are doped into the semiconductor resulting in the formation of patchy *insulating* domains. Assume percolative electrical transport occurs through the pristine domains when channel resistance is measured from source to drain in a transistor device. Derive a relationship between the fraction of insulating domains and the resistance of the channel. State all assumptions. Note that percolative electrical transport is simply given by $\frac{\sigma}{\sigma_m} = (p - p_c)^l$, where σ is conductivity, σ_m is conductivity of the pristine phase, p is the fraction of pristine state, p_c is percolation threshold, and l depends on dimensionality, $l = 2$ for 3-D percolation.

19. Consider ionic drift under pulsed electric fields. Using the Nernst–Einstein relation as the starting point, estimate a simple relation between the drift distance for an ionic species under an applied voltage in a two-terminal memristor channel.

20. Estimate the majority carrier response time for n-doped silicon (for nominal doping concentration of $10^{15}/cm^3$). For simplicity, assume a one-dimensional case. The dielectric constant of silicon can be taken to be 12.

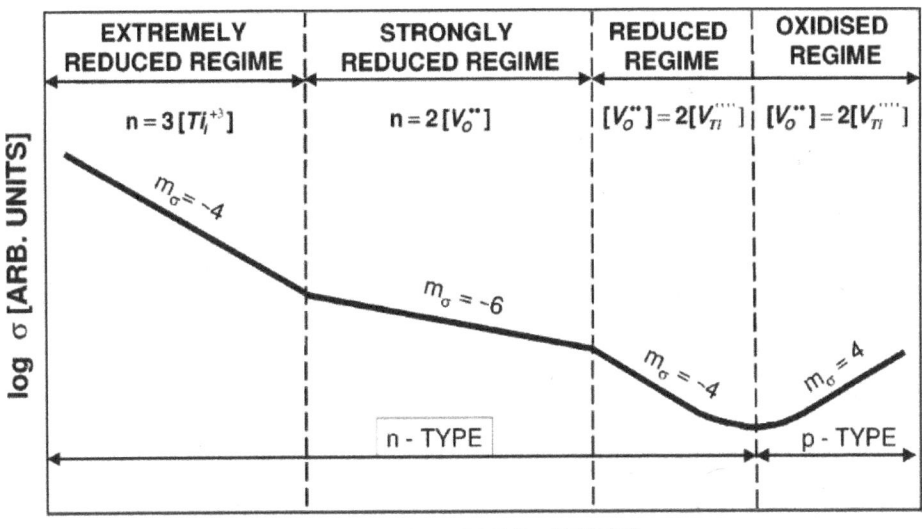

Figure 9.2 Defect chemical diagram for titanium dioxide as function of oxygen partial pressure. Adapted with permission from [2]. Copyright © 2006 American Chemical Society.

21. An important model semiconductor being studied for memristor technologies and resistive memory is TiO_2. The electrical conductivity of TiO_2 is extremely sensitive to the defect density due to different mechanisms that are onset at different regimes of non-stoichiometry. Graphically, we can represent the dominant defects through a Brouwer diagram as shown in Figure 9.2.

Derive thermodynamic relationships for point defect equilibria for each of the four regimes marked in Figure 9.2 and prove that the dependence of conductivity on oxygen activity is distinct for each major defect type. Note that in a memristor device, the equivalent of reducing and oxidizing regimes are realized by controlling the voltage bias polarity to form and migrate oxygen vacancies across from an electrode interface. This contributes to the SET, RESET operations as well as analog modulation of resistance.

22. In the case of thermal runaway mechanism, often there is a square-root relationship between threshold voltage for abrupt resistance switching and ambient temperature. Let us assume we have predominantly PF conduction in the artificial neuron candidate material of interest. A sharp rise in current flow occurs when thermal energy balances effective reduction in energy barrier at a critical field. Derive a functional relationship between threshold voltage and ambient temperature to explain the observed experimental trend.

23. Although ferroelectric polarization is typically nonvolatile, in nanoscale materials, the depolarization field can be large enough to spontaneously reduce the net polarization. For the case of a bilayer stack comprising a dielectric–ferroelectric heterostructure, explain what happens to the depolarization field as the dielectric thickness is varied. How will this parameter influence the design of a ferroelectric synapse?

24. Magnetic memory devices can be fabricated from tunnel junctions wherein the probability of electron tunneling across two ferromagnetic layers separated by a thin insulator is dependent on the relative magnetization orientation. Assume f and f' are the fractions of tunneling electrons whose moments are parallel to the magnetization in the two magnetic layers, respectively. Assume that the temperature is low enough such that electrons tunnel across an insulating barrier without spin flips. Derive a relation for change in conductance when the magnetization is parallel versus antiparallel for the two magnetic layers. This is the well-known Jullière model of tunneling.

25. Domain walls are important features in magnetic materials that host a variety of interesting properties. If the two main factors controlling the width of a domain wall are Heisenberg exchange energy and crystal anisotropy energy, derive an expression for the width in terms of spin quantum number S, lattice

constant a, and Heisenberg exchange coefficient J_{ex}. Assume a simple cubic lattice for this problem.

26. Consider a single layered neural network where the neurons have lateral inhibitory recurrent connections. Let us consider that the membrane potential $u(t)$ of a particular neuron is represented by: $u(t+1) = u(t) + B - \sum_i w_i I_i$, where B represents a constant external input to the network at each timestep (can encode the weighted sum of pixel intensities from the input layer) and w_i represents the lateral inhibitory weights driven by the spikes I_i of neurons in that layer. The network operates in an event-driven fashion where the events are spikes of the neurons. In a CMOS implementation, there will be energy overhead to update the membrane potential of the neurons even if there are no spikes at a particular timestep. Design a time-hopping strategy where the network directly hops to the next timestep a neuron will fire. Compensate for the error in the membrane potential update of the neurons in the network by accounting for the skipped timesteps.

27. Revisit the above problem by considering a leak factor in the membrane potential update equation.

28. Derive Eq. (6.1) in Chapter 6.

29. Derive Eq. (6.5) using Eqs. (6.2) and (6.4) noted in Chapter 6.

30. Train a fully connected neural network on the MNIST dataset with 1 hidden layer consisting of 100 neurons. Report the training convergence plot and the final testing error. Use the ReLU transfer function. Use all constraints for ANN–SNN conversion (no bias, no batch normalization, use average pooling). Now convert the ANN to an SNN. You will need to now simulate the ANN model over a number of timesteps, replace the ReLUs with IF-spiking neurons, and convert the images to a Poisson spike train (use a spike rate of 100% for the highest pixel intensity). Report the converted SNN accuracy along with the plot of accuracy versus timesteps.

31. Using the SNN trained in the previous problem, vary the maximum firing rate of the input. What is the optimal firing rate that minimizes the energy consumption? Use the following proxy metric for evaluating energy consumption: For every spike from a fan-in input neuron, an accumulate operation will take place at the receiving neuron. Report energy consumption in terms of total number of accumulate operations. Since accuracy will vary as a function of timesteps, use an accuracy value close enough to your baseline ANN accuracy for determining optimal rate (i.e., perform iso-accuracy comparison). Justify your analysis, that is, explain and show graphs to substantiate the trade-offs involved in varying the firing rate.

32. Using the network trained in the previous problem, include synaptic device constraints in the network. Incorporate the following constraints in your synaptic weights: resistance ON–OFF ratio (10) and discrete number of resistance states (4 bits – 16 states). Optimize to reach near ideal software based SNN accuracy. Explain your optimization process and how you included the constraint in your code. Report accuracies after you included each constraint.
33. Train a neural network with 1 hidden layer of 500 neurons using the following learning rules: BPTT, DECOLLE, and e-prop on the N-MNIST and TIMIT datasets. Report the testing accuracies.
34. Add recurrent connections in the SNN architecture for the previous problem and evaluate the testing accuracies for both the datasets. Comment on any changes in the testing accuracies.
35. Consider training a VGG-16 SNN architecture on the ImageNet dataset using a hybrid approach where the network is first converted from an ANN and then fine-tuned using BPTT learning algorithm. Do you observe any reduction in the inference latency of the SNN?

References

[1] Zhou, Y. and Ramanathan, S., 2012. Relaxation dynamics of ionic liquid – VO_2 interfaces and influence in electric double-layer transistors. *Journal of Applied Physics*, *111*(8), pp. 084508-1–084508-7.
[2] Nowotny, M. K., Bak, T. and Nowotny, J., 2006. Electrical properties and defect chemistry of TiO_2 single crystal. I. Electrical conductivity. *The Journal of Physical Chemistry B*, *110*(33), pp. 16270–16282.

Index

2D semiconductors, 64

action potentials, 5
activation function, 20
address event representation, 115
adversarial, 13
analog CMOS, 78
analog neural networks, 28
antiferromagnet, 68
Aplysia, 10
ASCII, 45
associative learning, 10
astrocytes, 139
asynchronous, 114
axon hillock, 6

backend of line, 38
backpropagation, 21
back-propagation through time (BPTT), 148
band gap, 87
Bayesian, 137
binary classification, 15
binary neural networks, 137
bit precision, 31
Boolean summation, 101

C. elegans, 11, 176
catastrophic forgetting, 155
chemical synapses, 4
ciliate, 10
circadian, 9
classical conditioning, 10
computer vision, 15
convolutional neural networks, 15
correlated oxides, 91
crossbar arrays, 87
Curie temperature, 38
current oscillations, 95

decision nodes, 15
Deep Continuous Local Learning (DECOLLE), 153
deep learning, 16

deep neural networks, 25
dendritic action potentials (dCaAPs), 172
dendritic spines, 7
dendrocentric, 169
depolarization-induced suppression
    of excitation (DSE), 139
depolarizing field, 89
depression, 41
diffusional, 4
direct feedback alignment, 153
domain wall, 67
Dropout, 24

electrochromic, 47
electrolytic gates, 125
electron-doped, 122
eligibility trace, 4, 151
endocannabinoids mediated synaptic
    potentiation (e-SP), 139
endurance, 33
epitaxial, 41
e-prop, 151
equilibrium potential, 6
equilibrium propagation (EP), 158
event-driven computation, 28
exchange bias, 68
excitatory input, 6

FeFET, 90
ferroelectric, 38
ferroelectric tunnel junctions, 39, 87
filamentary, 33
filamentary pathways, 85
free layer, 66

glial, 5
gradient descent, 17
grain boundaries, 105
Grow when Required (GWR), 155

habituation, 9
Hebbian plasticity, 3

hidden–output, 18
Hodgkin–Huxley, 85
homeostasis, 80
Hopfield networks, 159

IGZO, 81
implicit differentiation
    at equilibrium (IDE), 158
in vivo, 180
In-Memory, 113
In-Memory computing, 31
input–hidden, 18
insulator–metal transition, 46
interneurons, 11
ion channels, 5

Joule heating, 86

Kirchhoff's law, 31
Knowledge Distillation, 162
Kullback–Leibler (KL) divergence, 137

Lapicque model, 178
large language models (LLMs), 15, 158
lateral spin valve, 105
layered semiconductors, 104
leaky-integrate and fire, 77
lifelong learning, 169
linear regression, 126
lithography, 33

maximum likelihood criterion, 155
memory access, 30
memory leakage, 30
mono-domain, 105
monolayer graphene, 104
monolithic, 122
Moore's law, 13
Mott insulators, 98
Multiply and Accumulate Unit, 30
myelin sheaths, 118

nanomagnetic, 65
negative differential resistance (NDR), 172
Nernst equation, 6
nervous system, 2
neurogenesis, 155
neuromodulators, 4
neuronal cells, 2
Nodes of Ranvier, 6
non-associative, 11
nonvolatile, 31
notches, 68

on-chip learning, 32
operant conditioning, 10
organic semiconductor, 98
orthorhombic, 41

oscillatory neural networks, 119
oxygen vacancies, 81

paired-pulse facilitation, 51
parasitics, 32
path optimization, 83
pattern classification, 121
percolation, 85
perovskite nickelates, 121
phase-change memory, 108
phase delay, 121
phase inhomogeneity, 122
pinned layer, 66
polarization domains, 87
polyaniline, 124
polyethylene terephthalate, 99
polymers, 43
Poole–Frenkel, 84
postsynaptic, 3
potentiation, 41
power consumption, 13
presynaptic, 3
probabilistic neurons, 105

reactional, 4
rectified linear unit, 19, 28
recurrent neural network (RNN), 148
refractory period, 6, 80
regression, 16
regularization, 24
reservoir computing, 126
reset by subtraction, 77
resistive divider, 105
retention, 32
ring resonator, 71

scanning tunneling microscope, 64
Schottky barrier, 36
Schottky diode, 82
self-oscillations, 84
Sense Amplifiers, 129
sensitization, 9
sensory neurons, 11
short-term plasticity, 56
sigmoid, 19
skyrmion, 69
slime mold, 10
sparsity, 114
spatial memory, 1
spike-timing-dependent
    plasticity (STDP), 145
spiking neural networks, 28
spin Hall effect, 67
Static Random Access Memory, 30
straight-through estimator (STE), 155
structural distortions, 105
STT-MRAM, 66
subthreshold, 79

superparamagnetic, 108
supervised learning, 15
synaptogenesis, 155

thermal noise, 105
threshold switching, 95
transfer function, 105
transformer, 15
transient memory, 120

transmission probability, 140
tunnel current, 91

Van der Waals semiconductors, 65
variability, 32
voltage controlled magnetic anisotropy, 67

waveguide, 70
winner-takeall (WTA), 145

Printed by Integrated Books International,
United States of America